U.S.NRC
United States Nuclear Regulatory Commission
Protecting People and the Environment

I0489267

FISCAL YEAR 2011
PERFORMANCE AND ACCOUNTABILITY REPORT

MISSION

License and regulate the Nation's civilian use of byproduct, source, and special nuclear materials to ensure adequate protection of public health and safety, promote the common defense and security, and protect the environment.

TABLE OF CONTENTS

This Performance and Accountability Report is available on the NRC Website http://www.nrc.gov

Left to right: Commissioner William D. Magwood IV, Commissioner Kristine L. Svinicki, Chairman Gregory B. Jaczko, Commissioner George Apostolakis, and Commissioner William C. Ostendorff.

The FY 2011 Performance and Accountability Report provides performance results and audited financial statements that enable Congress, the President, and the public to assess the performance of the agency in achieving its mission and stewardship of its resources. The report contains a concise overview, Management's Discussion and Analysis, as well as performance and financial sections. Details of performance results and program evaluations can be found in the Other Accompanying Information section.

I am pleased to present the U.S. Nuclear Regulatory Commission's (NRC's) Performance and Accountability Report (PAR) for Fiscal Year (FY) 2011. The report provides key financial and performance information to Congress and the American people. For the tenth consecutive year, we received the Certificate of Excellence in Accountability Reporting from the Association of Government Accountants for our FY 2010 PAR. This prestigious award recognizes our commitment to accountability and high-quality reporting of performance and financial information. This report highlights our achievements and challenges in meeting our mission through the agency's two strategic goals of safety and security.

In FY 2011, we continued to maintain effective and efficient oversight of the Nation's 104 nuclear reactors, placing continued emphasis on strengthening the interrelationship among safety, security, and emergency preparedness. We continued to review all safety aspects of new reactor designs, environmental siting, and combined license applications for the construction of new nuclear power plants. We also remained focused on the safe and secure use of nuclear materials through effective oversight of fuel facilities, uranium recovery sites, decommissioning sites, and nuclear material user licensees.

The devastating earthquake and tsunami that struck Japan on March 11 led to what is now widely recognized as one of the worst accidents in the history of nuclear power. In the aftermath of the Fukushima Dai-ichi emergency, the NRC took strong and immediate actions to ensure the continued safety of our Nation's nuclear power plants. Throughout our response, the NRC staff exhibited the same high level of dedication and professionalism that I have seen consistently throughout my seven years with the Commission.

The Commission established the Near-Term Japan Task Force, comprised of some of the agency's most senior staff and experts, to conduct a very thorough review of all available information and to develop a comprehensive set of recommendations for strengthening nuclear safety. Their work was made possible by the assistance of hundreds of other NRC staff who supported their efforts. This has been a tremendous accomplishment for the Task Force and the agency as a whole. The Japan situation required NRC staff to address new safety challenges while continuing to maintain a focus on our other important responsibilities. It also resulted in a high level of congressional and public interest in our response to the Fukushima accident and how we will continue to ensure nuclear safety in this country.

We are committed to prudently managing the resources entrusted to us by the American people. We will continue to evaluate, test, and strengthen our internal controls, including those related to financial reporting and financial management systems, as required by the *Federal Managers' Financial Integrity Act of 1982* (FMFIA). Based on the FMFIA assessments, I have concluded that there is reasonable assurance that the NRC is in substantial compliance with FMFIA, and the financial and performance data published in this report is complete, accurate, reliable, and timely, in accordance with the *Reports Consolidation Act of 2000* and Office of Management and Budget Circular A-136 requirements. Additionally, I have determined that the agency is in substantial compliance with the *Federal Financial Management Improvement Act of 1996* (FFMIA), based on the NRC's application of the FFMIA risk model.

I am proud of the performance of NRC employees in achieving the agency's Safety and Security goals and look forward to continuing the high-quality service the American people have come to expect from us.

Gregory B. Jaczko

Gregory B. Jaczko
Chairman
November 9, 2011

CERTIFICATE OF EXCELLENCE IN ACCOUNTABILITY REPORTING®

BEST-IN-CLASS AWARD

Presented to the

U.S. Nuclear Regulatory Commission

In recognition for Providing the

Most Comprehensive and Candid
Presentation of Forward-Looking Information

in your FY10 Performance and Accountability Report

John R. Hummel, CGFM
Chair, Certificate of Excellence
in Accountability Reporting Board

Relmond P. Van Danker, DBA, CPA
Executive Director, AGA

CERTIFICATE OF EXCELLENCE IN ACCOUNTABILITY REPORTING®

Presented to the

U.S. Nuclear Regulatory Commission

In recognition of your outstanding efforts
preparing NRC's Performance and
Accountability Report for the fiscal
year ended September 30, 2010.

A Certificate of Excellence in Accountability Reporting is presented
by AGA to federal government agencies whose annual
Performance and Accountability Reports achieve the
highest standards demonstrating accountability
and communicating results.

John H. Hummel, CGFM
Chair, Certificate of Excellence
in Accountability Reporting Board

Relmond P. Van Danker, DBA, CPA
Executive Director, AGA

Chapter 1

MANAGEMENT'S DISCUSSION AND ANALYSIS

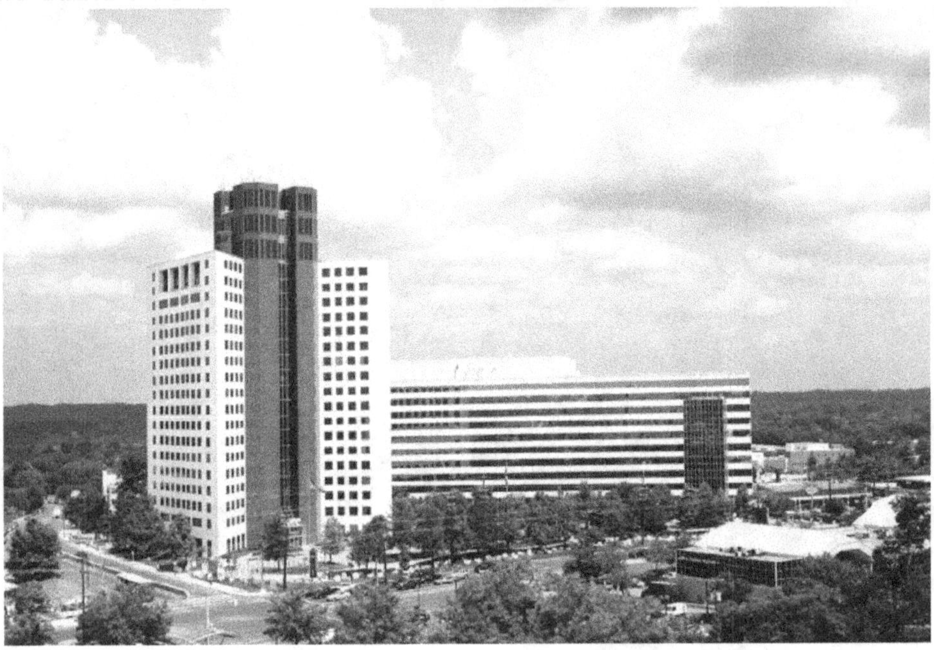

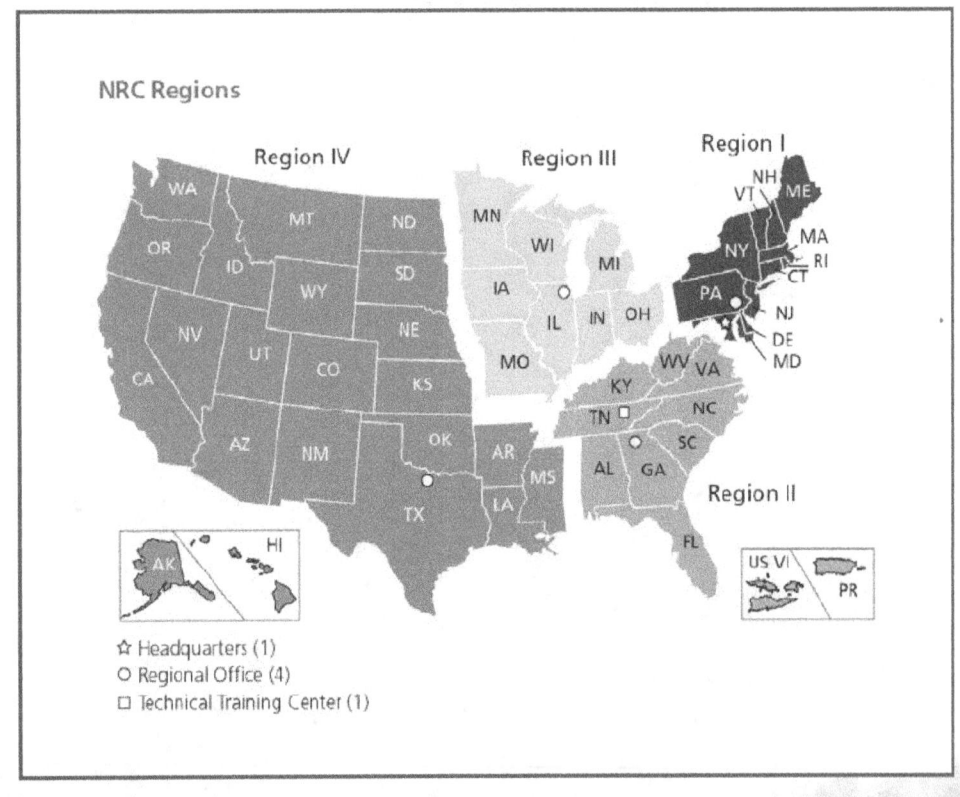

The U.S. Nuclear Regulatory Commission (NRC) Headquarters

NRC Regions
Region IV
Region III
Region I
Region II
WA
OR
ID
MT
ND
SD
WY
NE
NV
UT
CO
CA
KS
AZ
NM
OK
TX
AR
LA
MS
MN
WI
MI
IA
IL
IN
OH
MO
NH
VT
ME
MA
RI
CT
NY
PA
NJ
DE
MD
WV
VA
KY
TN
NC
SC
AL
GA
FL
AK
HI
US VI
PR
☆ Headquarters (1)
○ Regional Office (4)
□ Technical Training Center (1)

Chapter 1
MANAGEMENT'S DISCUSSION AND ANALYSIS

INTRODUCTION

The U.S. Nuclear Regulatory Commission (NRC) Performance and Accountability Report is an account of the agency's effectiveness and efficiency in achieving its mission. This report describes the agency's program and financial management performance during Fiscal Year (FY) 2011, which covers the period from October 1, 2010, to September 30, 2011.

The NRC had a very successful year in FY 2011. The agency has two strategic goals: Safety and Security. The agency achieved both its Safety and Security goals and met 10 of 11 performance measure targets. The agency also improved its operational activities by continuing to invest in its skilled workforce of engineers and scientists through knowledge transfer programs, recruiting a diverse workforce, and providing training opportunities.

The NRC was recognized for the third consecutive year as the *Best Place to Work in the Federal Government* by the non-profit organization, Partnership for Public Service (PPS), and was ranked the second most innovative Federal agency by PPS and the Hay Group, a global management consulting firm.

The NRC is in a sound financial position, having sufficient funds to meet programmatic needs and adequate control of these funds in place. The agency also received an unqualified audit opinion on its financial statements by independent auditors, with no instances of noncompliance with laws and regulations.

This report consists of four chapters. Chapter 1, "Management's Discussion and Analysis," provides an overview of the NRC and describes its programmatic and financial accomplishments during FY 2011. Chapter 2, "Program Performance," describes the agency's success in meeting its goals and describes the programmatic activities that are the basis for accomplishing those goals.

Chapter 3, "Financial Statements and Auditor's Report," details the agency's financial position. Chapter 4, "Other Accompanying Information," includes information on management challenges, a summary of the financial statement audit, and other information. The NRC places a high priority on keeping the public informed of its activities. Please visit our Website http://www.nrc.gov to access this report and to learn more about who we are and what we do to serve the American public.

ABOUT THE NRC

The NRC is an independent Federal agency. The *Atomic Energy Act of 1954*, as amended, and the *Energy Reorganization Act of 1974*, as amended, define the NRC's purpose. These acts provide the foundation for the NRC's mission that regulates the Nation's civilian use of byproduct, source, and special nuclear materials in order to protect public health and safety, to promote the common defense and security, and to protect the environment. The agency regulates civilian nuclear power plants, other nuclear facilities, and other uses of nuclear materials. These other uses include nuclear medicine programs at hospitals; academic activities at educational institutions; research work; industrial applications, such as gauges and testing equipment; and the transport, storage, and disposal of nuclear materials and wastes.

The NRC is headed by a Commission composed of five members, with one member designated by the President to serve as Chairman. With the advice and consent of the Senate, the President appoints each member to serve a 5-year term. The Chairman is the principal executive officer and official spokesman for the Commission. The Executive Director for Operations carries out policies and decisions made by the Commission and directs the activities of the programs.

The NRC's headquarters is located in Rockville, MD. The NRC has an Operations Center housed within the

Commissioner

William D. Magwood, IV

Commissioner

Kristine L. Svinicki

Executive Director,
Advisory Committee on
Reactor Safeguards

Edwin M. Hackett

Chief Administrative
Judge (Chairman),
Atomic Safety and
Licensing Board Panel

E. Roy Hawkens

Director, Office of
Commission Appellate
Adjudication

Brooke D. Poole

Director, Office of
Congressional Affairs

Rebecca L. Schmidt

Director,
Office of Public Affairs

Eliot B. Brenner

Deputy Executive Director
for Reactor and
Preparedness Programs

Martin J. Virgilio

Regional
Administrator
Region I

Bill Dean

Regional
Administrator
Region II

Victor McCree

Regional
Administrator
Region III

Cynthia D. Pederson
(Acting)

Regional
Administrator
Region IV

Elmo E. Collins

Director,
Office of
New Reactors

Michael R. Johnson

Director, Office of
Nuclear Security and
Incident Response

James T. Wiggins

Director, Office of
Nuclear Reactor
Regulation

Eric J. Leeds

Director, Office
of Small Business
and Civil Rights

Corenthis B. Kelley

The dotted line signifies that the Inspector General exercises a much higher degree of independence with the Chairman in carrying out his roles and responsibilities in comparison to other executives reporting to the Chairman.

COMMISSION ORGANIZATIONAL CHART

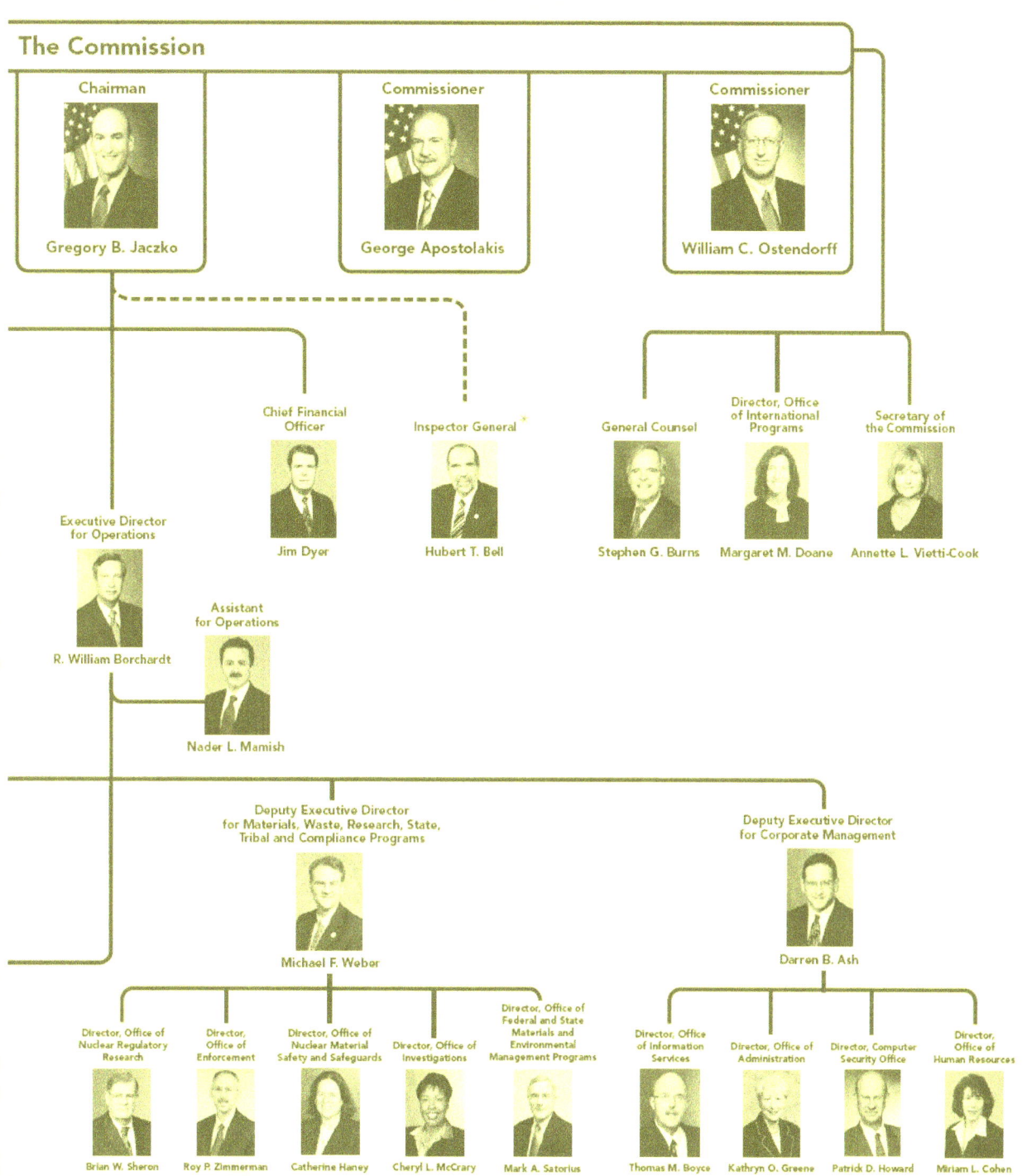

The Commission

Chairman
Gregory B. Jaczko

Commissioner
George Apostolakis

Commissioner
William C. Ostendorff

Chief Financial Officer
Jim Dyer

Inspector General *
Hubert T. Bell

General Counsel
Stephen G. Burns

Director, Office of International Programs
Margaret M. Doane

Secretary of the Commission
Annette L. Vietti-Cook

Executive Director for Operations
R. William Borchardt

Assistant for Operations
Nader L. Mamish

Deputy Executive Director for Materials, Waste, Research, State, Tribal and Compliance Programs
Michael F. Weber

Deputy Executive Director for Corporate Management
Darren B. Ash

Director, Office of Nuclear Regulatory Research
Brian W. Sheron

Director, Office of Enforcement
Roy P. Zimmerman

Director, Office of Nuclear Material Safety and Safeguards
Catherine Haney

Director, Office of Investigations
Cheryl L. McCrary

Director, Office of Federal and State Materials and Environmental Management Programs
Mark A. Satorius

Director, Office of Information Services
Thomas M. Boyce

Director, Office of Administration
Kathryn O. Greene

Director, Computer Security Office
Patrick D. Howard

Director, Office of Human Resources
Miriam L. Cohen

headquarters complex that coordinates communications with its licensees, State agencies, and other Federal agencies. This center is the focal point for assessing and responding to operating events in the industry. NRC operations officers staff the Operations Center 24 hours a day, seven days a week.

The NRC also has a regional office located in King of Prussia, PA; Atlanta, GA; Lisle, IL; and Arlington, TX. The regional offices allow the agency to work closely with the agency's licensees to ensure safety. The NRC also employs at least two resident inspectors at each of the Nation's nuclear power reactor sites.

The NRC's budget for FY 2011 was $1,054.2 million, with 3,991 full-time equivalent staff. The NRC budget is primarily covered by fees assessed to its licensees and applicants for a license. The agency collected approximately 90 percent of its budget from licensees, and applications for licenses, with the remaining funding provided by the U.S. Treasury.

THE NRC'S REGULATORY ACTIVITIES

To fulfill its responsibility to protect public health and safety, the NRC performs five principle regulatory functions: developing regulations and guidance for applicants and licensees; licensing or certifying applicants to use nuclear materials, operate nuclear facilities, and decommissioning facilities; inspecting and assessing licensee operations and facilities to ensure that licensees apply with NRC requirements and taking appropriate follow-up or enforcement actions when necessary; evaluating operational experience of license facilities and activities; and conducting research, holding hearings, and obtaining independent reviews to support regulatory decisions (see Figure 1).

The standards and regulations established by the agency set the rules that users of radioactive materials must follow. Drawing upon the knowledge and experience of the agency's scientists and engineers, these rules are the

basis for protecting workers and the general public from the potential hazards of radioactive materials.

With a few exceptions, any organization or individual intending to have or use radioactive materials must obtain a license. A license identifies the type and amount of radioactive material that may be held and used. NRC scientists and engineers evaluate the license application to ensure that the potential licensee's use of nuclear materials meets the agency's safety and security requirements.

Figure 1
HOW WE REGULATE

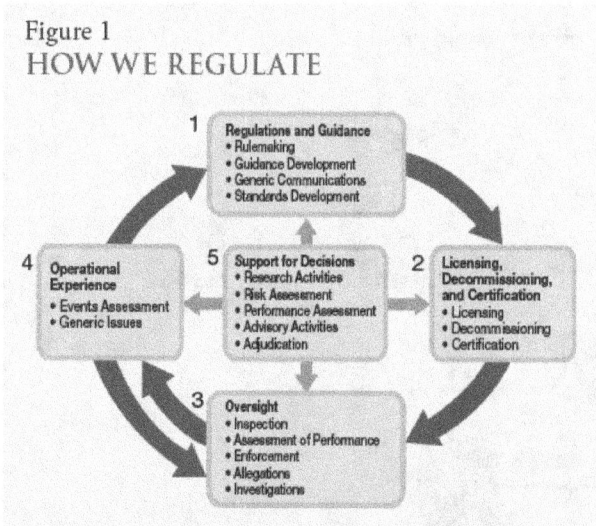

On a regular basis, the NRC inspects all facilities that it licenses to ensure that they meet agency regulations and are being operated safely and securely. NRC specialists conduct 10 to 25 routine inspections each year at each of the 104 operating nuclear power plants. In addition, the agency oversees 3,000 licenses for medical, academic, industrial, and general uses of nuclear materials. The agency conducts approximately 1,200 health and safety inspections of its nuclear materials licensees annually. Under the NRC's Agreement State program, 37 States have assumed primary regulatory responsibility over the industrial, medical, and other users of nuclear materials in their States, accounting for 19,200 licensees. The NRC works closely with these States to ensure that they maintain public safety through acceptable licensing and inspection procedures.

The NRC evaluates operational experience of licensed facilities and activities as a means to confirm that safety is being maintained. The NRC initiated the Industry Trends Program to monitor trends in indicators of industry performance. Should any long-term indicators show a statistically significant adverse trend, the NRC will evaluate them and take appropriate regulatory action using its processes for resolving generic issues and by issuing generic communications to licensees.

Finally, the NRC uses a number of independent sources to review and verify the quality of NRC regulatory decisions. This involves research activities, risk assessments, advisory groups, and adjudicatory hearings.

THE NUCLEAR INDUSTRY

The NRC is responsible for regulating all aspects of the civilian nuclear industry. The industry can best be described by examining the nuclear material cycle. The nuclear material cycle begins with the mining and production of nuclear fuel, continues with the use of nuclear fuel to power the Nation's 104 nuclear power plants, and ends with the safe transportation and storage of spent nuclear fuel and other nuclear waste. The agency's regulatory programs ensure that radioactive materials are used safely and securely at every stage in the nuclear material cycle. To address safety and security issues, the NRC has developed regulatory practices, knowledge, and expertise specific to each activity in the nuclear material cycle.

FUEL FACILITIES

The production of nuclear fuel begins at uranium mines, where milled uranium ore is used to produce a uranium concentrate called "yellowcake." At a special facility, the yellowcake is converted into uranium hexafluoride (UF_6) gas and loaded into cylinders. The cylinders are sent to a gaseous diffusion plant, where uranium is enriched for use as reactor fuel. The enriched uranium is then converted into oxide powder, fabricated into fuel pellets (each about

the size of a fingertip), loaded into metal fuel rods about 3.5 meters long, and bundled into reactor fuel assemblies at a fuel fabrication facility. Assemblies are then transported to nuclear power plants, non-power research reactor facilities, and naval propulsion reactors for use as fuel (see Figure 2). The NRC licenses eight major fuel fabrication and production facilities and three enrichment facilities in the United States (see Figure 3). Because they handle extremely hazardous material, these facilities take special precautions to prevent theft, diversion by terrorists, and dangerous exposures to workers and the public from this nuclear material.

Figure 2
SIMPLIFIED FUEL FABRICATION PROCESS

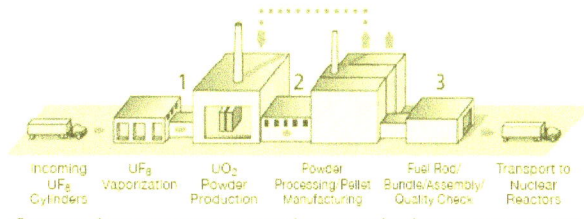

Incoming UF_6 Cylinders — UF_6 Vaporization — UO_2 Powder Production — Powder Processing/Pellet Manufacturing — Fuel Rod/Bundle/Assembly/Quality Check — Transport to Nuclear Reactors

Fabrication of commercial light-water reactor fuel consists of the following three basic steps:
(1) the chemical conversion of UF_6 to uranium dioxide (UO_2) powder
(2) a ceramic process that converts UO_2 powder to small ceramic pellets
(3) a mechanical process that loads the fuel pellets into rods and constructs finished fuel assemblies

Figure 3
LOCATIONS OF FUEL CYCLE FACILITIES

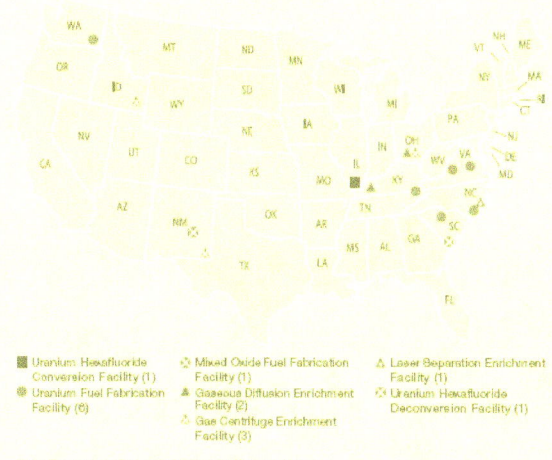

■ Uranium Hexafluoride Conversion Facility (1)
◉ Uranium Fuel Fabrication Facility (6)
◇ Mixed Oxide Fuel Fabrication Facility (1)
▲ Gaseous Diffusion Enrichment Facility (2)
△ Gas Centrifuge Enrichment Facility (3)
△ Laser Separation Enrichment Facility (1)
◇ Uranium Hexafluoride Deconversion Facility (1)

Note: There are no fuel cycle facilities in Alaska or Hawaii.

REACTORS

Power plants change one form of energy into another. Electrical generating plants convert heat energy, the kinetic energy of wind or falling water, or solar energy into electricity. A nuclear power plant converts heat energy into electricity. Other types of heat-conversion plants burn coal, oil, or gas to produce heat energy that is then used to produce electricity. Nuclear energy cannot be seen. There is no burning of fuel in the usual sense. Rather, energy is given off by the nuclear fuel as certain types of atoms split in a process called nuclear fission. This energy is in the form of fast-moving particles and invisible radiation. As the particles and radiation move through the fuel and surrounding water, the energy is converted into heat. The radiation energy can be hazardous, and facilities take special precautions to protect people and the environment from these hazards.

Because the fission reaction produces potentially hazardous radioactive materials, nuclear power plants are equipped with safety systems to protect workers, the public, and the environment. Radioactive materials require careful use because they produce radiation, a form of energy that can damage human cells. Depending on the amount and duration of the exposure, radiation

NRC Employee Conducting an Inspection at Byron Nuclear Plant

can potentially cause cancer. In a nuclear reactor, most hazardous radioactive substances, called fission byproducts, are trapped in the fuel pellets or in the sealed metal tubes holding the fuel. However, small amounts of these radioactive fission byproducts, principally gases, become mixed with the water passing through the reactor. Other impurities in the water also become radioactive as they pass through the reactor. The facility processes and filters the water to remove these radioactive impurities and then returns the water to the reactor cooling system.

MATERIALS USERS

The medical, academic, and industrial fields all use nuclear materials. For example, about one-third of all patients admitted to U.S. hospitals are diagnosed or treated using radioisotopes. Most major hospitals have specific departments dedicated to nuclear medicine. In all, about 112 million nuclear medicine or radiation therapy procedures are performed annually, with the vast majority used in diagnoses. Radioactive materials used as a diagnostic tool can identify the status of a disease and minimize the need for surgery. Radioisotopes give doctors the ability to look inside the body and observe soft tissues and organs, in a manner similar to the way x-rays provide images of bones. Radioisotopes carried in the blood also allow doctors to detect clogged arteries or check the functioning of the circulatory system.

The same property that makes radiation hazardous can also make it useful in treating certain diseases like cancer. When living tissue is exposed to high levels of radiation, cells can be destroyed or damaged. Doctors can selectively expose cancerous cells (cells that are dividing uncontrollably) to radiation to either destroy or damage these cells.

Many of today's industrial processes also use nuclear materials. High-tech methods that ensure the quality of manufactured products often rely on radiation generated

Chapter 1
MANAGEMENT'S DISCUSSION AND ANALYSIS

by radioisotopes. To determine whether a well drilled deep into the ground has the potential for producing oil, geologists use nuclear well-logging, a technique that employs radiation from a radioisotope inside the well to detect the presence of different materials. Radioisotopes are also used to sterilize instruments; find flaws in critical steel parts and welds that go into automobiles and modern buildings; authenticate valuable works of art; and solve crimes by spotting trace elements of poison. Radioisotopes can also eliminate dust from film and compact discs and reduce static electricity (which may create a fire hazard) from can labels. In manufacturing, radiation can change the characteristics of materials, often giving them features that are highly desirable. For example, wood and plastic composites treated with gamma radiation resist abrasion and require low maintenance. As a result, they are used for some flooring in high-traffic areas of department stores, airports, hotels, and churches.

WASTE DISPOSAL

Typically, a nuclear power plant generates the following two types of radioactive waste: high-level waste, which consists of used fuel (usually called spent fuel), and low-level waste, which includes contaminated equipment, filters, maintenance materials, and resins used in purifying water for the reactor cooling system. Other users of radioactive materials also generate low-level waste.

Nuclear power plants handle each type of radioactive waste differently. They must use special procedures in the handling of the spent fuel because it contains the highly radioactive fission byproducts created while the reactor was operating. The spent fuel from nuclear power plants is stored in water-filled pools at each reactor site or at a storage facility in Illinois. The water in the spent fuel storage pool provides cooling and adequately shields and protects workers from the radiation. Several nuclear power plants also use dry casks to store spent fuel. These heavy metal or concrete casks rest on concrete pads adjacent to the reactor facility. The thick layers of concrete and steel in these casks shield workers and the public from radiation.

In FY 2008, the Department of Energy (DOE) submitted and the NRC docketed for review, an application for a disposal facility at Yucca Mountain, NV. Subsequently, in FY 2010, the DOE filed a motion to withdraw its application with prejudice. In January 2010, Secretary of Energy Steven Chu convened the Blue Ribbon Commission on America's Nuclear Future to conduct a comprehensive review of policies for managing the back end of the nuclear fuel cycle and recommend a new plan. In the interim, most spent fuel in the U.S. will remain safely stored at individual plants.

Licensees often store low-level waste on site until its radioactivity has decayed and the waste can be disposed of as ordinary trash, or until amounts are large enough for shipment to a low-level waste disposal site in containers approved by the U.S. Department of Transportation. The NRC has developed a waste classification system for low-level radioactive waste based on its potential hazards and has specified disposal and waste form requirements for each of the following general classes of waste: Class A, Class B, and Class C waste. Generally, Class A waste contains lower concentrations of radioactive material than Class B and Class C wastes. There are two low-level disposal facilities that accept a broad range of low-level wastes. They are located in Barnwell, SC, and Richland, WA.

FY 2011 PERFORMANCE RESULTS

The NRC's Strategic Plan describes the agency's mission, goals, and strategies. The Strategic Plan can be found at the NRC Website http://www.nrc.gov. The agency's two strategic goals are focused on safety and security. The Safety goal is to *Ensure adequate protection of public health and safety and the environment*. The Security goal is to *Ensure adequate protection in the secure use and management of radioactive materials*.

Safety is the primary goal of the NRC. The agency achieves this goal by ensuring that the performance of licensees is at or above acceptable safety levels. NRC safety programs work in conjunction with the agency's licensees in a partnership. The NRC licensees are responsible for designing, constructing, and operating nuclear facilities safely. The NRC is responsible for regulatory oversight of the licensees. The NRC Safety goal activities are designed to achieve the following strategic outcomes:

STRATEGIC OUTCOMES

- Prevent the occurrence of any nuclear reactor accidents.
- Prevent the occurrence of any inadvertent criticality events.
- Prevent the occurrence of any acute radiation exposures resulting in fatalities.
- Prevent the occurrence of any releases of radioactive materials that result in significant radiation exposures.
- Prevent the occurrence of any releases of radioactive materials that cause significant adverse environmental impacts.

These strategic outcomes specify the conditions under which the Safety goal can be considered to have been met.

SAFETY GOAL STRATEGIES

The NRC used the following safety strategies from its Strategic Plan to guide its activities and to achieve its Safety goal in FY 2011:

1. Develop, maintain, and implement licensing and regulatory programs for reactors, fuel facilities, materials users, spent fuel management, uranium recovery, and decommissioning activities to ensure the adequate protection of public health and safety, and the environment.

2. Continue to oversee the safe operation of existing plants while preparing for and managing the review of applications for new power reactors.

3. Conduct the NRC safety, security, and emergency preparedness programs in an integrated manner.

4. Improve the NRC's regulatory programs and apply safety-focused research to anticipate and resolve safety issues.

5. Use sound science and state-of-the-art methods to establish, where appropriate, risk-informed and performance-based regulations.

6. Promote focused attention on safety matters and individual accountability of those engaged in regulated activities.

7. Utilize domestic and international operating experience to inform decision-making.

8. Oversee licensee safety performance through inspections, investigations, enforcement, and performance assessment activities.

9. Effectively respond to events at NRC-licensed facilities and other events of national interest, including maintaining and enhancing the NRC's critical incident response and communication capabilities.

FY 2011 RESULTS

In FY 2011, the NRC achieved all five of its Safety goal strategic outcomes. The NRC also uses six performance measures to determine whether it has met its Safety goal. The agency met all six performance measure targets in FY 2011 (see Table 1).

The first three performance measures focus on performance at individual nuclear power plants. Inspection results show that all of the nuclear power plants are operating safely.

Clinton Nuclear Generating Station

The fourth measure tracks the trends of several key indicators of nuclear power plant safety. This measure is the broadest measure of the safety of nuclear power plants, incorporating the performance results from all plants to determine industry average results. This measure shows that there were no statistically significant adverse trends in any of the indicators in FY 2011.

The last two safety performance measures track harmful radiation exposures to the public and occupational workers, and radiation exposures that harm the environment. Neither of these two measures exceeded their targets in FY 2011.

Table 1: SAFETY PERFORMANCE MEASURE SCORECARD

Safety Performance Measures	2006	2007	2008	2009	2010	2011
1. Number of new conditions evaluated as red by the Reactor Oversight Process is ≤ 3.	0	0	0	0	0	1
2. Number of significant accident sequence precursors of a nuclear reactor accident is zero.	0	0	0	0	0	0
3. Number of operating reactors with integrated performance that entered the Inspection Manual Chapter 0350 process, or the multiple/repetitive degraded cornerstone column, or the unacceptable performance column of the Reactor Oversight Process Action Matrix, with no performance exceeding Abnormal Occurrence Criterion I.D.4 is ≤ 3.	0	1	0	0	0	2
4. Number of significant adverse trends in industry safety performance with no trend exceeding the Abnormal Occurrence Criterion I.D.4 is ≤ 1.	0	0	0	0	0	0
5. Number of events with radiation exposures to the public and occupational workers that exceed Abnormal Occurrence Criterion I.A. is:						
▪ Reactors: 0	0	0	0	0	0	0
▪ Materials: ≤ 2	0	0	0	0	0	0
▪ Waste: 0	0	0	0	0	0	0
6. Number of radiological releases to the environment that exceed applicable regulatory limits is:						
▪ Reactor: 0	0	0	0	0	0	0
▪ Materials: ≤ 2	0	0	0	0	0	0
▪ Waste: 0	0	0	0	0	0	0

Chapter 1
MANAGEMENT'S DISCUSSION AND ANALYSIS

FUTURE CHALLENGES

The industry has experienced a substantial improvement in safety at nuclear power plants over the past 36 years as both the nuclear industry and the NRC have gained substantial experience in the operation and maintenance of nuclear power facilities. Despite its excellent safety record, the agency faces key challenges, such as ensuring that the new generation of nuclear power plants are built and operated safely, and ensuring the safe disposal of nuclear waste. The NRC will continue to identify and address management challenges it faces in accomplishing its mission. The NRC's Inspector General has identified its most serious management and performance challenges facing the agency. These challenges are discussed in Chapter 4 of this report.

Japanese Earthquake Evaluation

The NRC established a special task force in FY 2011 to conduct a review of its processes and regulations in light of the events at the Fukushima Dai-ichi Nuclear Power Plant in Japan. The NRC reviewed the manner in which it requires licensees to protect nuclear power plants from natural disasters. The review found that, based on the current regulations and nuclear plant capabilities, a sequence of events like the Fukushima accident is unlikely to occur in the United States. Therefore, continued operation and licensing activities do not pose an imminent risk to public health and safety. However, the review did yield important insights into ways that the agency can improve its regulatory processes to account for events that exceed the current design-basis for natural disasters at nuclear power plants. The agency has reviewed the task force recommendations within the context of the NRC's existing regulatory framework and considered various regulatory vehicles available to the NRC to implement the recommendations and will proceed based on Commission direction.

Recommendations for Enhancing Reactor Safety in the 21st Century – The Near-Term Task Force Review of Insights from the Fukushima Dai-ichi Accident (http://pbadupws.nrc.gov/docs/ML1118/ML111861807.pdf)

Licensing New Reactors

Many factors contribute to the growing interest in nuclear power, such as rising electricity demand, clean-air concerns, performance and reliability of existing plants, and a wide range of policies included in the *Energy Policy Act of 2005* to encourage new reactor construction. In addition, the loan guarantee program, nuclear energy production tax credits for the first 6,000 megawatts of electricity from new advanced reactors, and standby insurance underwritten by the Federal Government have influenced the renewed interest in nuclear power.

Currently, the electric industry is pursuing plans to build 20 reactors based on five standard designs. As of June 2011, the NRC is reviewing 12 Combined Operating Licenses (COL) applications and expects to complete the licensing process for two applications in 2012. In an effort to improve regulatory efficiency and add greater predictability to the process, the NRC established *Title 10 of the Code of Federal Regulations* (10 CFR) Part 52, "Licenses Certification, Approvals for Nuclear Power Plants" that includes the issuance of a COL. The NRC approval is necessary before a nuclear power plant can be built and operated. The NRC maintains oversight of the construction and operation of a facility throughout its lifetime to ensure compliance with the Commission's regulations for the protection of public health and safety, the common defense and security, and the environment.

In addition to the interest in building new light-water reactors, there is also interest in small and simpler units for generating electricity from nuclear power. This interest is driven by a number of factors, including reducing capital costs, shortening overall construction periods, reducing carbon footprints by replacing older coal-fired generators, and providing dedicated and reliable power to critical infrastructure. The technologies that are being proposed are very diverse, although current interests are focusing on smaller Pressurized Water Reactors. Still, these new designs involve unique features and the vendors are proposing unique operating staffing models which present challenges to the NRC. The NRC is developing an appropriate regulatory framework for licensing and construction of these new designs. The NRC is also keeping abreast of the Department of Energy's Next Generation Nuclear Plant (NGNP) Demonstration Project and changes to the Department's plans for that project. The rapidly evolving nature of the so-called small module reactors, coupled with industry's expanding interest in having the NRC review design applications to support plans to build these new, small units in the next 7-10 years, presents the NRC with additional challenges. This will require the NRC to be flexible in adjusting review priorities based on applicant schedules and staff expertise available to conduct pre-application, safety, and environmental reviews in an effective and timely manner.

Integrated Spent Fuel Management

The ending of the Yucca Mountain high-level waste repository program will leave the Nation with a dilemma. Although the spent fuel remains safe and secure at more than 100 nuclear power plant sites, not to mention the spent fuel and high-level waste already being stored by DOE at other locations around the country, the amount of spent fuel and high-level waste continues to grow with each passing month. All commercially viable nuclear fuel cycles contemplate the need for some permanent disposal capacity. The Blue Ribbon Commission on America's Future is finalizing its assessment and recommendations for the Nation's path forward. The administration will consider the Commission's results as it develops a new strategy for nuclear waste management and disposal.

Although the amounts of waste continue to increase, one constant is the NRC's continued focus on ensuring safety, security, and environmental protection. This mission needs to be accomplished regardless of the uncertainties and the variables that exist. In response to the recent changes in the national program for high-level waste management, the NRC has initiated a number of actions, including the following:

■ evaluation of the technical and regulatory requirements to support long-term dry storage of spent fuel and deferred transportation of spent fuel

■ identification of regulatory gaps and development of a regulatory framework for reprocessing

■ consideration of a revised waste confidence rulemaking

■ development of revisions to NRC's regulatory and analytical tools to consider alternative waste disposal options.

Enhancing NRC Effectiveness and Efficiency in Business Support Services

Predicting the level of funding for the NRC's programs over the next several years is difficult. However, key factors suggest that the agency is entering an era of no growth or declining budgets. For the past two years, the NRC's budget has remained essentially flat. Further, with concerns over rising Federal spending and debt, there is strong congressional interest in reducing Government spending. With this fiscal outlook, it is essential for the NRC to find ways to improve the delivery of business support services, including (1) administrative services, (2) human capital, (3) financial management (including contract management), (4) information management (IM), and (5) information technology (IT) in a more effective and efficient way, thereby reducing operational costs. Consequently, the agency undertook a study entitled "Transforming Assets into Business Solutions (TABS)," to identify ways to optimize business processes, eliminate work that is no longer necessary, and reduce duplication and overlap. The TABS report presented 10 recommendations that provide the NRC with opportunities to centralize and standardize business support processes. The agency is currently implementing these recommendations to optimize the delivery of services and achieve cost savings over the next several years. It will be important to have these changes made in a manner that avoids disruption and does not compromise our ability to carry out our safety and security missions effectively.

STRATEGIC GOAL 2: SECURITY

Ensure Adequate Protection in the Secure Use and Management of Radioactive Materials

The NRC must remain vigilant in ensuring the security of nuclear facilities and materials in an elevated threat environment. The agency achieves its common defense and Security goal using licensing and oversight programs similar to those employed in achieving its Safety goal.

STRATEGIC OUTCOME

■ Prevent any instances where licensed radioactive materials are used domestically in a manner hostile to the security of the United States.

This strategic outcome specifies the condition under which the Security goal can be considered to have been met.

SECURITY GOAL STRATEGIES

The agency used the following security strategies from its Strategic Plan to guide its activities and achieve its Security goal in FY 2011:

1. Use relevant intelligence information and security assessments to maintain realistic and effective security requirements and mitigation measures.

2. Share security information with appropriate stakeholders and international partners.

3. Oversee licensee security performance through inspections and force-on-force exercises.

4. Control the handling and storage of sensitive security information and the communication of information to licensees and Federal, State, and local partners.

5. Support Federal response plans that employ an approach to the security of nuclear facilities and radioactive material that integrates the efforts of licensees and Federal, State, local, and Tribal governments.

6. Use a risk-informed approach to implement appropriate regulatory controls for the possession, handling, import, export, and transshipment of radioactive materials.

7. Enhance the programs for control of the security of radioactive sources and strategic special nuclear material commensurate with their risk, including enhancements required by the *Energy Policy Act of 2005*.

8. Promote U.S. national security interests and nuclear non-proliferation policy objectives for NRC-licensed imports and exports of source and special nuclear materials and nuclear equipment.

FY 2011 RESULTS

In FY 2011, the NRC achieved its Security goal strategic outcome. The NRC also uses five Security goal performance measures to determine whether the agency has met its Security goal. The agency met four of the five performance measure targets in FY 2011 (see Table 2).

The first performance measure tracks unrecovered losses or thefts of risk-significant radioactive sources. The measure ensures that those radioactive sources that the agency has determined to be risk-significant to the public health and safety are accounted for at all times. The ability to account for these sources is critical to secure the Nation from "dirty bomb" attacks or other means of radiation dispersal.

There were no losses and one theft of radioactive nuclear material that the NRC considered to be risk-significant during FY 2011. On July 19, 2011, in Austin, TX, the Licensee (Acuren Inspections, Inc.) notified the Texas Department of Health that a truck had been broken into and that a radiography camera transportation container holding a QSA Global Model 880 D camera with a 33.7-currie iridium (Ir)-192 source and a portable electric generator had been stolen. The agency will coordinate and review the increased controls applied to these sources and determine if additional controls need to be implemented for these sources. If changes to the increased controls are needed, they will also be considered in the ongoing Rulemaking for 10 CFR Part 37, *Physical Protections of Byproduct Material*.

The second, third, and fourth performance measures evaluate the number of significant security events and incidents that occur at NRC-licensed facilities. These measures determine whether nuclear facilities maintain adequate protective forces to prevent theft or diversion of nuclear material or sabotage; whether systems in place at licensee plants accurately account for the type and amount of materials processed, utilized, or stored; and whether the facilities account for special nuclear material at all times with no losses of this material. There were no events that met the conditions for these measures in FY 2011.

The last security measure tracks significant unauthorized disclosures of classified or safeguards information that may cause damage to national security or public safety. This measure focuses on whether classified information or safeguards information is stored and utilized in such a way as to prevent its disclosure to the public, terrorist organizations, other nations, or personnel without a need to know. Unauthorized disclosures can harm national security or compromise public health and safety. The measure also focuses on whether controls are in place to maintain and secure the various devices and systems (electronic or paper-based) that the agency and its licensees use to store, transmit, and utilize this information. There were no documented disclosures of this type of information during FY 2011.

Table 2: SECURITY PERFORMANCE MEASURE SCORECARD

Security Performance Measures	2006	2007	2008	2009	2010	2011
1. Number of unrecovered losses or thefts of risk-significant radioactive sources is zero.	0	0	0	0	0	1
2. Number of substantiated cases of theft or diversion of licensed, risk-significant radioactive sources or formula quantities of special nuclear material, or attacks that result in radiological sabotage, is zero.	0	0	0	0	0	0
3. Number of substantiated losses of formula quantities of special nuclear material or substantiated inventory discrepancies of formula quantities of special nuclear material that are caused by theft or diversion or by substantial breakdown of the accountability system sabotage is zero.	0	0	0	0	0	0
4. Number of substantial breakdowns of physical security or material control (i.e., access control containment or accountability system) that significantly weaken the protection against theft, diversion, or sabotage is less than or equal to one.	0	0	0	0	0	0
5. Number of significant, unauthorized disclosures of classified or safeguards information is zero.	0	0	0	0	0	0

SECURITY AT NUCLEAR FACILITIES

Nuclear facilities are among the most secure facilities in the Nation. The NRC, in concert with other Federal agencies, constantly monitors intelligence to determine the level of threat faced by nuclear facilities. The agency continues to improve the regulatory requirement to better ensure the security of nuclear materials and facilities because the threat faced from those seeking to steal classified information has become more urgent in recent years. Nuclear facilities have implemented increased security measures, including "force-on-force" training exercises, to help ensure protection of this vital national infrastructure.

The NRC has also focused on security concerns related to radioactive sources typically employed by radiation medicine and other non-power applications of nuclear technology. The sheer number of radioactive sources–numbering thousands in the U.S. alone–creates challenges in securing these sources. Moreover, these sources are widely spread geographically and used for a wide array of purposes. The agency will continue to evaluate ways to enhance its ability to account for these sources.

Finally, many nations around the world have demonstrated an interest in developing and expanding their use of peaceful applications of nuclear technology. The agency works across a broad range of international organizations, such as the International Atomic Energy Agency (IAEA), to provide assistance to these countries to put in place measures to focus attention on key security issues. As the world's largest nuclear regulatory authority, the NRC takes a leadership role in extending this type of assistance. The agency anticipates that its assistance to other countries will continue to promote the secure use of nuclear materials.

THE INDUSTRY TRENDS PROGRAM

In addition to its annual performance measures, the NRC gauges the effectiveness of its Nuclear Reactor Safety program based on its Industry Trends program. The NRC and stakeholder indicators of industry performance are a means to confirm that the safety of operating power plants is being maintained. The NRC compiles data on overall safety performance using several industry-level performance indicators, a number of which are described below. These indicators show significant improvement in the long-term trends for safety performance of nuclear power plants. Plant operating experience data have shown a steady stream of improvements in the reliability of plant systems and components, plant operating procedures, training of power plant operators, and regulatory oversight. For ease of viewing, all the charts in this section display data since 1993.

The industry safety indicators are derived through engineering and scientific analyses by the agency. The performance indicator results are subject to minor variations as licensees submit revisions to the source data and may differ slightly from data reported in previous years as a result of refinements in data quality. The results of these analyses are reported annually to both the Commission and Congress.

Significant Events

Significant events meet specific criteria such as degradation of important safety equipment. The agency reviews operating events and assesses their safety significance. The number of significant events has followed a declining statistical trend.

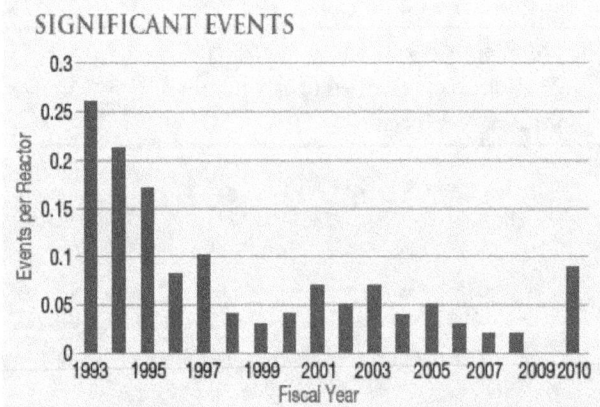

SIGNIFICANT EVENTS

Chapter 1
MANAGEMENT'S DISCUSSION AND ANALYSIS

Radiation Exposure

The total (collective) radiation dose received by workers is an indication of the radiological challenges of maintaining and operating nuclear power plants. The trend shows a reduction in collective dose and demonstrates the effectiveness of the controls on radiation exposure implemented to meet these challenges.

Automatic Scrams

A scram is a basic reactor protection safety function that shuts down the reactor by inserting control rods into the reactor core. Scrams can result from events that range from relatively minor incidents to precursors of accidents. The massive power blackout in August 2003 accounts for most of the increase in FY 2003, but it has not affected the statistical trend for number of scrams, which has been declining steadily.

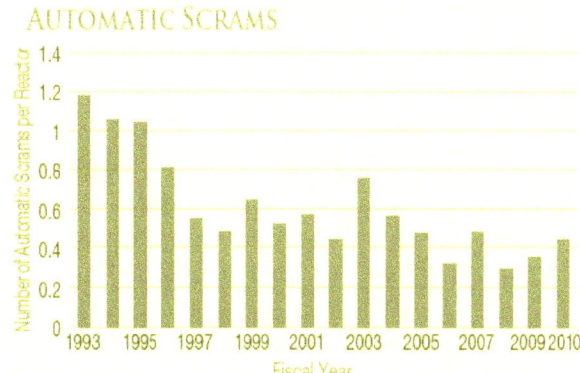

Safety System Actuations

Safety systems mitigate off-normal events, such as the widespread power blackout in August 2003, by providing reactor core cooling and water addition. Actuations of safety systems that are monitored include certain emergency core cooling and emergency electrical power systems. Actuations can occur as a result of "false alarms" (such as testing errors) or in response to actual events.

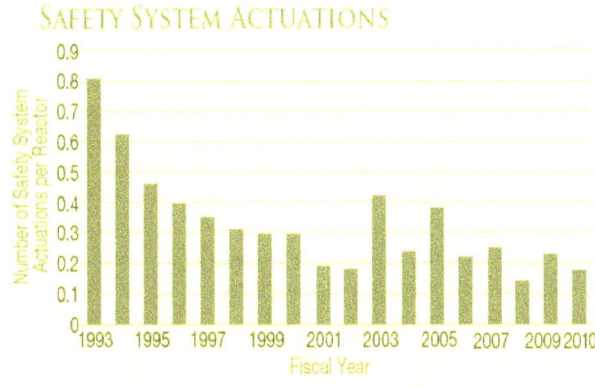

Precursor Occurrence Rate

A precursor event is an event that has a probability of greater than 1 in 1 million of leading to substantial damage to the reactor fuel. There is no statistically significant adverse trend in the occurrence rate of precursor events since 1993, the baseline year for the statistical analysis. In addition, no statistically significant trend is detected for all precursors during the FY 2001–2009 period. Due to the complexities associated with evaluating precursor events, the data always lag behind other indicators.

PRECURSOR OCCURRENCE RATE

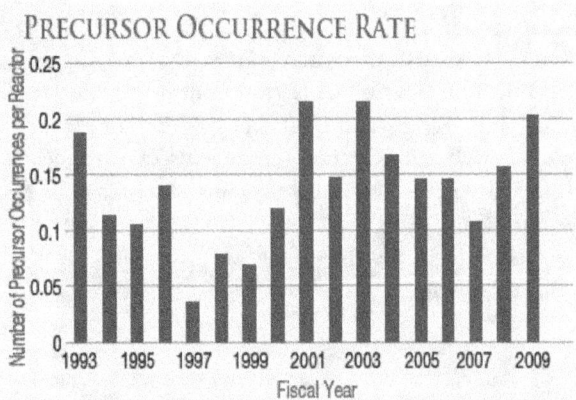

Safety System Failures

Safety system failures include any events or conditions that could prevent a safety system from fulfilling its safety function. The statistical trend for number of safety system failures across the industry has been declining.

SAFETY SYSTEM FAILURES

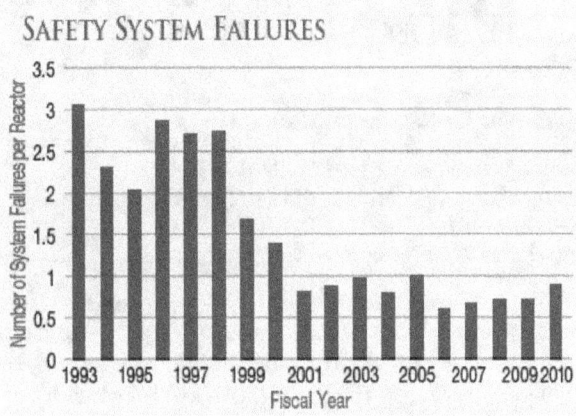

DATA COMPLETENESS AND RELIABILITY

The NRC considers the data contained in this report to be complete, reliable, and relevant. The data are complete because the agency reports actual performance data for every performance goal and indicator in the report. In addition, all of the data are reported for each measure. The agency also considers the data in this report reliable and relevant, because they have been

validated and verified. "Verification and Validation of NRC's Performance Measures and Metrics" contains the processes the agency uses to collect, validate, and verify performance data in this report. This report can be found in Appendix III of the NRC's FY 2011 Congressional Budget Justification located on the NRC Website http://www.nrc.gov/reading-rm/doc-collections/nuregs/staff/sr1100/v26/sr1100v26.pdf.

FINANCIAL PERFORMANCE OVERVIEW

As of September 30, 2011, the financial condition of the NRC was sound with respect to having sufficient funds to meet program needs, and adequate control of these funds was in place to ensure obligations did not exceed budget authority. The NRC prepared its financial statements in accordance with the accounting standards codified in the Statements of Federal Financial Accounting Standards (SFFAS) and Office of Management and Budget (OMB) Circular A-136, *Financial Reporting Requirements.*

SOURCES OF FUNDS

The NRC has two appropriations, Salaries and Expenses and Office of the Inspector General. Funds for both appropriations are available until expended. The NRC's new FY 2011 budget authority was $1,054.2 million, which was reduced by a $0.3 million rescission of funds, bringing the total new budget authority to $1,053.9 million. Of this amount, $1,043.1 million was for the Salary and Expenses appropriation of which $9.9 million was derived from the Nuclear Waste Fund for activities relating to the *Nuclear Waste Policy Act of 1982* (NWPA), as amended, and $10.8 million was for the Office of the Inspector General appropriation. This represents a decrease in new budget authority of $13.0 million compared to FY 2010 [$13.0 million for the Salaries and Expenses appropriation and no change for the Office of the Inspector General appropriation] (see Figure 4). In addition, $52.6 million from prior-year appropriations, $10.9 million from prior-year reimbursable work, and $14.5 million for new reimbursable work to be performed for others was available to obligate in FY 2011. The sum of all funds available to obligate for FY 2011 was $1,131.9 million, which was a $31.9 million decrease from the FY 2010 amount of $1,163.8 million.

The *Omnibus Budget Reconciliation Act of 1990*, as amended, requires the NRC to collect fees to offset approximately 90 percent of its new budget authority, less the amount appropriated to the NRC from the Nuclear Waste Fund, amounts appropriated for waste incidental to reprocessing and generic homeland security for FY 2011. The projected amount to be received from reactor and material fees in FY 2011 was $916.2 million after accounting for billing adjustments. The NRC collected $911.0 million of the required $915.8 million in fees for the year, which was 99.5 percent of the 90 percent fee recovery requirement.

Figure 4
SOURCES OF FUNDS (PROJECTED)

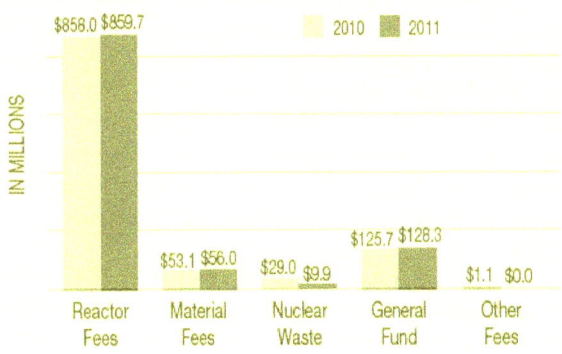

USES OF FUNDS BY FUNCTION

The NRC incurred obligations of $1,083.5 million in FY 2011, which was a decrease of $35.6 million over FY 2010 (see Figure 5). Approximately 58 percent of obligations were used for salaries and benefits. The remaining 42 percent was used to obtain technical assistance for the NRC's principal regulatory programs, to conduct confirmatory safety research, to cover operating expenses (e.g., building rentals, transportation, printing,

security services, supplies, office automation, training), to pay for staff travel, and to cover reimbursable work.

The unobligated budget authority available at the end of FY 2011 was $48.5 million, a $3.8 million increase compared to the FY 2010 amount of $44.7 million. Of this $48.5 million, $13.1 million was for reimbursable work and $35.4 million was available to fund critical NRC needs in FY 2012.

Figure 5
USES OF FUNDS BY FUNCTION

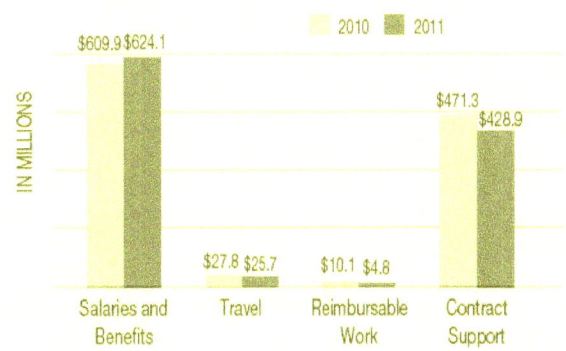

AUDIT RESULTS

The NRC received an unqualified audit opinion on its FY 2011 financial statements and internal controls. The auditors found no instances of noncompliance or substantial noncompliance with laws and regulations during the FY 2011 audit.

A summary of the financial statement audit results is included in the "Other Accompanying Information" section of this report.

LIMITATIONS OF THE FINANCIAL STATEMENTS

The principal financial statements have been prepared to report the financial position and results of operations of the NRC, pursuant to the requirements of 31 U.S.C. 3515(b). While the statements have been prepared from the books and records of the NRC in accordance with Generally Accepted Accounting Principles (GAAP) for Federal entities and the formats prescribed by OMB, the statements are in addition to the financial reports used to monitor and control budgetary resources, which are prepared from the same books and records. The statements should be read with the realization that they are for a component of the U.S. Government, a sovereign entity.

FINANCIAL STATEMENT HIGHLIGHTS

The NRC's financial statements summarize the financial activity and financial position of the agency. The financial statements, footnotes, and required supplementary information appear in Chapter 3. Analysis of the principal statements follows.

ANALYSIS OF THE BALANCE SHEET

ASSET SUMMARY (IN MILLIONS)

As of September 30,	2011	2010
Fund Balance with Treasury	$394.6	$420.1
Accounts Receivable, Net	100.3	130.9
Property & Equipment, Net	46.5	36.2
Other	3.7	3.1
Total Assets	$545.1	$590.3

Assets. The NRC's assets were $545.1 million as of September 30, 2011, a decrease of $45.2 million from the end of FY 2010. The decrease is primarily due to decreases of $30.6 million in Accounts Receivable, Net and $25.5 million in the Fund Balance with Treasury, offset by a $10.3 million increase in Property & Equipment, Net.

The Fund Balance with Treasury was $394.6 million at September 30, 2011, accounting for 72 percent of total assets. This account represents appropriated funds, collected license fees, and other funds maintained at the U.S. Department of the Treasury (Treasury) to pay current liabilities and to finance authorized purchase commitments. The $25.5 million decrease in the fund balance is primarily the result of a decrease of $28.5 million in the beginning balance compared with the prior year. Receipts from appropriated funds decreased $13.0 million from FY 2010 as a result of new budget authority (including a decrease of $19.1 million for the Nuclear Waste Fund and a $0.3 million rescission of FY 2011 current year funds returned to Treasury), offset by $18.0 million for the FY 2010 rescission of prior-year unobligated funds that were returned to Treasury resulting in a net increase of $5.0 million in the fund balance. Fees collected, and then transferred to Treasury, increased $1.4 million over FY 2010, having a net offsetting effect on the fund balance. (The revenue generated by fees assessed to licensees as required by law is sent to Treasury to offset approximately 90 percent of its appropriations received during the year.) Payments, which reduce the fund balance, had a net decrease of $0.3 million and comprised primarily of a decrease of $40.2 million in general disbursements, offset by increases of $36.3 million in salaries and benefits disbursements and $2.3 million in grant disbursements.

Accounts receivable consists of amounts owed to the NRC by other Federal agencies and the public. Accounts Receivable, Net, as of September 30, 2011, was $100.3 million, which includes an offsetting allowance for doubtful accounts of $4.5 million. This is a 23 percent decrease from the FY 2010 year-end Accounts Receivable, Net, balance of $130.9 million. The variance is due to a reduction of $30.6 million in accounts receivable for material and facilities fees for work performed for licensees, primarily as a result of a process change for invoicing inspection fees.

Chapter 1
MANAGEMENT'S DISCUSSION AND ANALYSIS

LIABILITIES SUMMARY (IN MILLIONS)

As of September 30,	2011	2010
Accounts Payable	$ 43.2	$ 40.5
Federal Employee Benefits	7.2	7.6
Other Liabilities	79.2	112.0
Total Liabilities	$129.6	$160.1

Liabilities. Total liabilities were $129.6 million as of September 30, 2011, a decrease of $30.5 million from the FY 2010 year-end balance of $160.1 million. The decrease is primarily due to a change in Other Liabilities of $32.8 million, resulting from the removal in FY 2011 of a Contingent Liability recorded in FY 2010 of $11.8 million for the probable likelihood of an adverse outcome of legal claims, and a decrease of $17.5 million in accrued funded salaries and benefits.

Liabilities not covered by budgetary resources were $58.9 million, representing 45 percent of total liabilities. The liabilities not covered by budgetary resources in FY 2011 include $49.9 million in unfunded accrued annual leave included in Other Liabilities for the amount of leave earned but not yet taken and $9.0 million in accrued and future workers' compensation. Compared to the prior year, liabilities not covered by budgetary resources showed a decrease of 18 percent from the balance of $71.5 million as of September 30, 2010. The decrease of $12.6 million is primarily due to a contingent liability recorded in FY 2010 of $11.8 million which was removed in FY 2011.

NET POSITION SUMMARY (IN MILLIONS)

As of September 30,	2011	2010
Unexpended Appropriations	$310.3	$311.9
Cumulative Results of Operations	105.2	118.3
Total Net Position	$415.5	$430.2

Net Position. The difference between Total Assets and Total Liabilities, Net Position, was $415.5 million as of September 30, 2011, a decrease of $14.7 million from the FY 2010 year-end balance. Net Position is comprised of two components: Unexpended Appropriations and Cumulative Results of Operations. Unexpended Appropriations is the amount of spending authority granted by Congress that remains unused by the agency. Unexpended Appropriations had a slight change from the prior fiscal year end at $310.3 million with a decrease of $1.6 million. Cumulative Results of Operations which represents the cumulative excess of financing sources over expenses, decreased $13.1 million.

ANALYSIS OF THE STATEMENT OF NET COST

Net costs are gross costs offset by earned revenue. The Statement of Net Cost presents the net cost of the NRC's two programs as identified in the NRC Annual Performance Plan. The purpose of this statement is to link program performance to the cost of programs. The NRC's Net Cost of Operations for the year ended September 30, 2011, was $208.2 million, which is a decrease of $8.8 million over the FY 2010 net cost of $217.0 million.

NET COST OF OPERATIONS (IN MILLIONS)

For the years ended September 30,	2011	2010
Nuclear Reactor Safety and Security	$ 70.8	$ 46.3
Nuclear Materials & Waste Safety and Security	137.4	170.7
Net Cost of Operations	$208.2	$217.0

The NRC's total gross costs decreased $43.6 million. Gross costs decreased $25.0 million in the Nuclear Reactor Safety and Security program primarily in the area of the Operating Reactor business line. The Nuclear

Materials & Waste Safety and Security program gross costs decreased $18.6 million primarily due to workload changes in the Nuclear Materials Users and High-Level Waste business lines, offset by increases in the Spent Fuel Storage and Transportation business line.

Total earned revenue at September 30, 2011, was $888.7 million, a decrease of $34.8 million from the earned revenue for the year ended September 30, 2010, which was $923.5 million. The decrease is primarily due to a reduction of $30.6 million in accounts receivable for material and facilities fees due to a process change for invoicing inspection fees.

Fees collected (earned primarily in FY 2011) and transferred to Treasury during FY 2011 were $911.0 million, compared to $909.5 million for FY 2010. The NRC is required to collect approximately 90 percent of appropriations for NRC activities through fee billing. Fees for reactor and materials licensing and inspections are collected in accordance with Title 10 of the *Code of Federal Regulations* (10 CFR) Part 170, "Fees for Facilities, Materials, Import and Export Licenses, and Other Regulatory Services under the *Atomic Energy Act of 1954*, as Amended," and 10 CFR Part 171, "Annual Fees for Reactor Licenses and Fuel Cycle Licenses and Materials Licenses, Including Holders of Certificates of Compliance, Registrations, and Quality Assurance Program Approvals and Government Agencies Licensed by the NRC."

ANALYSIS OF THE STATEMENT OF CHANGES IN NET POSITION

The Statement of Changes in Net Position reports the change in net position during the reporting period. Net position is affected by changes in its two components— Cumulative Results of Operations and Unexpended Appropriations. The decrease in Net Position of $14.7 million from FY 2010 to FY 2011 is due to decreases of $13.1 million in Cumulative Results of Operations and $1.6 million in Unexpended Appropriations.

The decrease in Cumulative Results of Operations of $13.1 million is primarily due to the decreases in the beginning balance of $10.1 million and in financing sources of $11.8 million, offset by $8.8 million in the net cost of operations. Financing sources primarily include inputed financing costs absorbed by others, and appropriations used which are funds consumed, reduced by the collection of fees assessed and the Nuclear Waste Funding expense. Imputed finance costs increased $9.8 million due to a cost recorded in FY 2011 of $12.7 million for judgements and awards, offset by a decrease of $2.9 million in cost for retirement and health benefits. Appropriations used decreased $2.5 million from the prior year primarily due to a decrease in funds consumed of $19.6 million, reduced by a decrease in the Nuclear Waste Funding expense of $18.6 million, offset by an increase in collection of fees assessed of $1.5 million.

A change in unexpended appropriations primarily results from appropriations received and adjustments (e.g., rescissions) being more, or less, than appropriations used during the fiscal year. In FY 2011, appropriations received of $133.3 million consisted of the NRC's total appropriation of $1,053.9 million (including a $0.3 million rescission for current year funds), reduced by $911.0 million in fee collections returned to Treasury and the Nuclear Waste Fund transfer of $9.9 million. Appropriations used in FY 2011 totaled $134.6 million and consisted of funds used of $1,060.2 million, reduced by collection from fees assessed of $911.0 million and Nuclear Waste Fund expenses of $14.6 million.

ANALYSIS OF THE STATEMENT OF BUDGETARY RESOURCES

The Statement of Budgetary Resources reports the source and status of budgetary resources at the end of the period.

Chapter 1
MANAGEMENT'S DISCUSSION AND ANALYSIS

It presents the relationship between budget authority and budget outlays, and the reconciliation of obligations to total outlays. For FY 2011, the NRC had Total Budgetary Resources available of $1,131.9 million, which is a decrease of $31.9 million from the $1,163.8 million available for FY 2010. Changes in budgetary resources include decreases of $13.0 million in the appropriation received during FY 2011 (which was $1,053.9 million, including a $0.3 million rescission for current year funds, compared to $1,066.9 million in FY 2010), $36.4 million due to a change in the beginning balance and $3.6 million due to recoveries of prior year unpaid obligations, offset by increases of $18.0 million due to a decrease in the rescission of prior year funds returned to Treasury and $3.1 million as a result of a change in spending authority from offsetting collections. The appropriation included decreases of $5.6 million for Nuclear Reactor Safety and Security, $7.4 million for Nuclear Materials and Waste Safety and Security, and no change for the Office of the Inspector General.

For FY 2011, the NRC had Obligations Incurred of $1,083.5 million, which is a decrease of $35.6 million from the FY 2010 Obligations Incurred of $1,119.1 million. Obligations Incurred had decreases of $22.5 million relating to the Nuclear Waste Fund, $8.3 million in NRC disbursements, and $5.3 million in reimbursable obligations, offset by an increase of $0.5 million for the Office of the Inspector General.

Gross outlays which represent funds disbursed during the year for current and prior year expenses, were $1,088.4 million for FY 2011, remaining basically the same as the FY 2010 gross outlays of $1,088.7 million. Major changes in outlay categories included a decrease in general disbursements (primarily for contract support) of $40.2 million, offset by increases in salary and benefits disbursements of $36.3 million, and grant disbursements of $2.3 million.

MANAGEMENT ASSURANCES, SYSTEMS, CONTROLS, AND LEGAL COMPLIANCE

This section provides information on NRC's compliance with the *Federal Managers' Financial Integrity Act of 1982* (Public Law 97-255), OMB Circular A-123, *Management's Responsibility for Internal Control*, and the *Federal Financial Management Improvement Act of 1996*.

FEDERAL MANAGERS' FINANCIAL INTEGRITY ACT

The *Federal Managers' Financial Integrity Act of 1982* (Integrity Act) mandates that agencies establish internal control to provide reasonable assurance that the agency complies with applicable laws and regulations; safeguards assets against waste, loss, unauthorized use, or misappropriation; and properly accounts for and records revenues and expenditures. The Integrity Act encompasses program, operational, and administrative areas, as well as accounting and financial management. It also requires the Chairman to provide an assurance statement on the adequacy of internal controls and on the conformance of financial systems with Government-wide standards, shown below.

INTERNAL CONTROL PROGRAM

Internal controls are the organization, policies, and procedures to help program and financial managers achieve results and safeguard the integrity of their programs. NRC managers are responsible for designing and implementing effective internal controls in their areas of responsibility. Each office director and regional administrator prepares an annual assurance certification that identifies any control weaknesses requiring the attention of the NRC Executive Committee on Internal Control (ECIC). These certifications are based on internal control activities such as risk assessments, as well as other activities such as Integrated Regulatory Review Service self-assessments, lessons learned oversight board activities, agency action review meetings, senior leadership meetings, audits of financial statements, reviews of financial statements, Inspector General and U.S. Government Accountability Office audits and reports, and other information provided by the congressional committees of jurisdiction.

The ECIC consists of senior executives from the Office of the Chief Financial Officer and the Office of the Executive Director for Operations. The agency's General Counsel and Inspector General participate as advisors.

The ECIC met and reviewed the reasonable assurance certifications provided by the offices and regions. The ECIC then informed the Chairman as to whether the NRC had any internal control deficiencies serious enough to require reporting as a weakness or noncompliance.

The NRC's internal control program requires that internal control deficiencies be documented and reported in office and regional internal control plans and operating plans. The internal control plans provide for annual reporting, and the operating plan process provides for quarterly updates; together, both ensure that key issues receive senior management attention. Combined with the individual assurance statements discussed previously, the internal control information in these plans provides the framework for monitoring and improving the agency's internal control on an ongoing basis.

U.S. NUCLEAR REGULATORY COMMISSION
FISCAL YEAR 2011
FEDERAL MANAGERS' FINANCIAL INTEGRITY ACT STATEMENT

The U.S. Nuclear Regulatory Commission (NRC) managers are responsible for establishing and maintaining effective internal control and financial management systems that meet the objectives of the *Federal Managers' Financial Integrity Act* (Integrity Act). The NRC conducted its assessment of internal control over programmatic operations in accordance with Office of Management and Budget (OMB) Circular A-123, *Management's Responsibility for Internal Control* (A-123) guidelines. Based on the results of this evaluation, NRC can provide reasonable assurance that its internal control over programmatic operations is in compliance with applicable laws and guidance, and no material weaknesses were found as of September 30, 2011.

In addition, NRC conducted its assessment of the effectiveness of internal control over financial reporting, which includes safeguarding of assets and compliance with applicable laws and regulations, in accordance with the requirements of Appendix A of A-123. Based on the results of the evaluation, NRC can provide reasonable assurance that its internal control over financial reporting as of June 30, 2011, was operating effectively, and no material weaknesses were found in the design or operation of the internal control over financial reporting.

The NRC can also provide reasonable assurance that its financial systems substantially comply with applicable Federal accounting standards as required by the *Federal Financial Management Improvement Act of 1996*.

Gregory B. Jaczko
Gregory B. Jaczko
Chairman
U.S. Nuclear Regulatory Commission
November 9, 2011

FY 2011 INTEGRITY ACT RESULTS

The NRC evaluated its internal control systems for the fiscal year ending September 30, 2011. Based on this evaluation, the NRC is able to provide a statement of assurance that the internal controls and financial systems meet the objectives of the Integrity Act. The NRC has reasonable assurance that its internal controls are effective and that its financial management systems conform to Government-wide standards.

Chapter 1
MANAGEMENT'S DISCUSSION AND ANALYSIS

OFFICE OF MANAGEMENT AND BUDGET CIRCULAR A-123, "MANAGEMENT'S RESPONSIBILITY FOR INTERNAL CONTROL"

Internal Control Over Financial Reporting (Appendix A)

In FY 2006, the NRC implemented the requirements of the revised OMB Circular A-123, which defined and strengthened management's responsibility for internal control in Federal agencies. The revised circular included updated internal control standards. Appendix A requires Federal agencies to assess the effectiveness of internal controls over financial reporting and to prepare a separate annual statement of assurance as of June 30, 2011.

In FY 2007, the NRC adopted a 3-year rotational testing plan. The agency determined that three of the original nine key processes were significant enough to include in the testing each year of the 3-year cycle. The remaining six key processes were to be tested once in the 3-year cycle, two each year. In FY 2011, the NRC continued its assessment of internal control over financial reporting. The agency reevaluated its scope of financial reports, materiality values, risk assessments, key processes, and key controls. Based on the results of this evaluation, the NRC can provide reasonable assurance that its internal control over financial reporting was operating effectively as of June 30, 2011, and the evaluation found no material weaknesses in design or operation of the internal controls over financial reporting.

Requirements For Effective Measurement And Remediation Of Improper Payments (Appendix C)

In FY 2011, OMB revised Parts I and II to Appendix C of OMB Circular A-123. Appendix C "Requirements for Effective Measurement and Remediation of Improper Payments," as amended, implemented the *Improper Payments Information Act of 2002* (IPIA) and the *Improper Payments Elimination and Recovery Act of 2010* (IPERA). The purpose of this guidance is to reduce improper payments, hold agencies accountable for reducing improper payments, and increase penalties for contractors who fail to disclose improper payments in a timely manner. The NRC complied with this guidance by incorporating improper payments testing into the FY 2011 A-123 Appendix A assessment.

FEDERAL FINANCIAL MANAGEMENT IMPROVEMENT ACT

The *Federal Financial Management Improvement Act of 1996* (FFMIA) requires each agency to implement and maintain systems that comply substantially with (1) Federal financial management system requirements, (2) applicable Federal accounting standards, and (3) the standard general ledger at the transaction level. FFMIA requires the Chairman to determine whether the agency's financial management systems comply with FFMIA and to develop remediation plans for systems that do not comply.

FY 2011 FFMIA RESULTS

As of September 30, 2011, the NRC evaluated its financial systems and found that they comply with applicable Federal requirements and accounting standards required by FFMIA. In making this determination, the agency considered all available information, including the report from the ECIC on the effectiveness of internal controls, Office of the Inspector General audit reports, and the results of the agency's financial management system reviews.

FINANCIAL MANAGEMENT SYSTEMS STRATEGIES

The NRC has started a business transformation initiative to develop an enterprise-wide financial system. The NRC plans to complete the business transformation in four distinct phases (or implementations). The four phases will cover the agency's core financial, acquisition, time and labor and budget formulation functions respectively. The objective is to consolidate and automate data and processes within a single integrated business solution to make the NRC a more transparent, efficient, and effective organization. During FY 2010, the first phase of our transformation was completed and five stand-alone legacy core financial systems were consolidated with nine subsystems into a new commercial-off-the-shelf core financial system (CFS). In FY 2012, the NRC upgraded its commercial-off-the-shelf Human Resources Management System (HRMS) for time and labor. The new HRMS system strengthened data security and introduced electronic workflow, eliminating paper and therefore reducing yearly costs. In FY 2014, the agency's acquisition function will be integrated with the CFS. After FY 2014, the NRC plans to complete our objective for an integrated and consolidated enterprise financial and acquisitions management system by consolidating the agency's time and labor, and budget formulation functions with the core financial and acquisitions functions.

PROMPT PAYMENT

The *Prompt Payment Act of 1982*, as amended, requires Federal agencies to make timely payments to vendors for supplies and services, to pay interest penalties when payments are made after the due date, and to take cash discounts when they are economically justified. In FY 2011, the NRC paid 88.9 percent of the 11,036 invoices subject to the Prompt Payment Act on time. The NRC did not meet its goal of 98 percent due to the deployment of a new accounting system and process changes. The NRC incurred $18,692 in interest penalties during FY 2011 (see Figure 6).

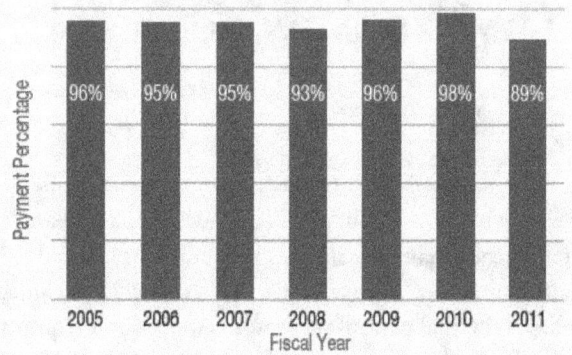

Figure 6
PROMPT PAYMENT

DEBT COLLECTION

The *Debt Collection Improvement Act of 1996* enhances the ability of the Federal Government to service and collect debts. The agency's goal is to maintain the level of delinquent debt owed to the NRC at year end to less than 1 percent of its annual billings. The NRC did not meet this goal, and at the end of FY 2011 delinquent debt was $13.7 million (see Figure 7). The NRC also failed to refer 100 percent of all eligible debt over 180 days delinquent to the U.S. Department of the Treasury for collection. These deficiencies are due to the deployment of a new accounting system and process changes. The NRC is taking steps to correct these deficiencies in FY 2012.

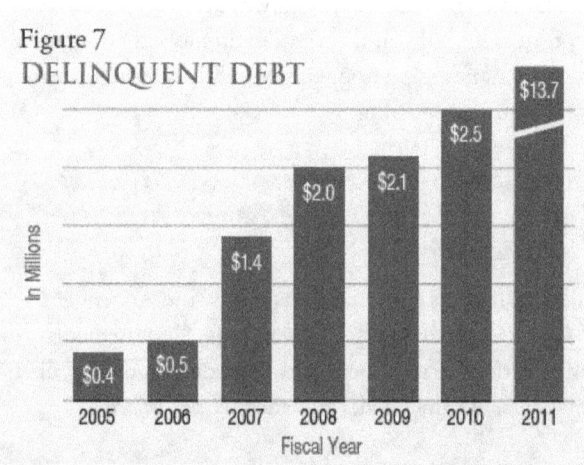

Figure 7
DELINQUENT DEBT

Chapter 1
MANAGEMENT'S DISCUSSION AND ANALYSIS

BIENNIAL REVIEW OF USER FEES

The *Chief Financial Officers Act of 1990* requires agencies to conduct a biennial review of fees, royalties, rents, and other charges imposed by agencies, and to make revisions to cover program and administrative costs incurred. Each year, the NRC revises the hourly rates for license and inspection fees and adjusts the annual fees to meet the fee collection requirements of the *Omnibus Budget Reconciliation Act of 1990*, as amended. The most recent changes to the license, inspection, and annual fees are described in the *Federal Register* (76 FR 36780, June 22, 2011).

The fees and charges for the Materials Access Authorization Program and Information Access Authority Program were also revised to more appropriately recognize actual costs. No other reviews were completed this year.

INSPECTOR GENERAL ACT OF 1978

The NRC has established and continues to maintain an excellent record in resolving and implementing Office of the Inspector General open audit recommendations. In the "Other Accompanying Information" section of this report, "Management Decisions and Final Actions on OIG Audit Recommendations," includes this information, as well as data concerning disallowed costs determined through contract audits conducted by the Defense Contract Audit Agency.

Chapter 2

PROGRAM PERFORMANCE

MEASURING AND REPORTING

This chapter presents detailed information on the NRC's performance in achieving its mission during FY 2011. It describes the NRC's performance results and program achievements in accomplishing its two strategic goals of safety and security.

The NRC's mission is to license and regulate the Nation's civilian use of byproduct, source, and special nuclear materials to ensure adequate protection of public health and safety, to promote the common defense and security, and to protect the environment.

Safety is the primary goal of the NRC. The agency's safety goal is to ensure adequate protection of public health and safety and the environment. The agency achieves this goal by ensuring that the performance of licensees is at or above acceptable safety levels. This chapter addresses the NRC's safety activities that regulate the agency's operating reactors, new reactors, fuel facilities, nuclear material users, decommissioning and low-level waste, spent fuel storage and transportation, and the proposed high-level waste repository licensees.

The agency's security goal is to ensure adequate protection in the secure use and management of radioactive materials. The NRC must remain vigilant in ensuring the security of nuclear facilities and materials in an elevated threat environment. The agency achieves its common defense and security goal using licensing and oversight programs for licensees similar to those employed in achieving its safety goal. The agency's security activities are also addressed in this chapter.

In addition, this chapter describes the agency's progress in achieving its organizational excellence objectives

of openness, effectiveness, and operational excellence. Finally, it describes information on data sources, data quality, and the completeness and reliability of performance data.

STRATEGIC GOAL 1: SAFETY
Ensure Adequate Protection of Public Health and Safety and the Environment

STRATEGIC OUTCOMES

The strategic outcomes specify the conditions under which an assessment can be made about whether the NRC has met its Safety goal. The NRC's Safety goal has five strategic outcomes that determine whether the agency has achieved its objective to ensure adequate protection of public health and safety and the environment:

- Prevent the occurrence of any nuclear reactor accidents.
- Prevent the occurrence of any inadvertent criticality events.
- Prevent the occurrence of any acute radiation exposures resulting in fatalities.
- Prevent the occurrence of any releases of radioactive materials that result in significant radiation exposures.
- Prevent the occurrence of any releases of radioactive materials that cause significant adverse environmental impacts.

In FY 2011, the NRC achieved all of its Safety goal strategic outcomes.

PERFORMANCE MEASURES

The NRC also uses annual performance measures to assess whether the agency met its Safety goal. Performance measures are aligned at a lower risk level than the strategic outcomes. As a result, not fully achieving a performance measure may not cause harm to the public or environment. Missing a performance

Chapter 2 PROGRAM PERFORMANCE

measure signals that safety levels may have deteriorated at the agency strategic planning level. If the NRC misses a performance measure, the agency will take corrective actions to bring the measure back into the target range. Table 3 below shows the agency's annual performance measures and results of FY 2006 – 2011.

Table 3: SAFETY GOAL PERFORMANCE MEASURES

Performance Measure	2006	2007	2008	2009	2010	2011
1. Number of new conditions evaluated as red by the Reactor Oversight Process is ≤ 3.	0	0	0	0	0	1
2. Number of significant accident sequence precursors of a nuclear reactor accident is zero.	0	0	0	0	0	0
3. Number of operating reactors with integrated performance that entered the Inspection Manual Chapter 0350 process, or the multiple/repetitive degraded cornerstone column, or the unacceptable performance column of the Reactor Oversight Process Action Matrix, with no performance exceeding Abnormal Occurrence Criterion I.D.4 is ≤ 3.	0	1	0	0	0	2
4. Number of significant adverse trends in industry safety performance with no trend exceeding the Abnormal Occurrence Criterion I.D.4 is ≤ 1.	0	0	0	0	0	0
5. Number of events with radiation exposures to the public and occupational workers that exceed Abnormal Occurrence Criterion I.A. is:						
▪ Reactors: 0	0	0	0	0	0	0
▪ Materials: ≤ 2	0	0	0	0	0	0
▪ Waste: 0	0	0	0	0	0	0
6. Number of radiological releases to the environment that exceed applicable regulatory limits is:						
▪ Reactor: ≤ 0	0	0	0	0	0	0
▪ Materials: ≤ 2	0	0	0	0	0	0
▪ Waste: 0	0	0	0	0	0	0

Chapter 2
PROGRAM PERFORMANCE

ANALYSIS OF FY 2011 PERFORMANCE MEASURE RESULTS

1. *Reactor Oversight Process*

The NRC reactor inspection program monitors nuclear power plant performance in three areas: (1) reactor safety, (2) radiation safety, and (3) security. Analysis of plant performance is based on many performance indicators and inspection findings. Each finding is then sorted into one of four categories: green, white, yellow, or red. Red indicates findings of high safety significance. There was one red finding for Browns Ferry Unit 1 in FY 2011.

2. *Reactor Significant Precursors*

This statistical measure of risk determines the likelihood of an event adversely impacting safety. A significant precursor is an event that has a probability of 1 in 1,000 (or greater) of leading to substantial damage to the reactor fuel. The NRC has identified no significant precursor events, based on screening reviews.

3. *Reactor Performance*

The conditions in this measure indicate whether the NRC finds significant performance issues in a plant during an inspection or based on performance indicators under the Reactor Oversight Process. If any of the conditions in this measure are met, the NRC will take action to ensure that plant safety is improved. Two reactors, Browns Ferry 1 and Fort Calhoun, were found to have met the conditions in this measure in FY 2011.

4. *Reactor Safety Trends*

This measure tracks trends for several key indicators of industry safety performance. These indicators provide insights into major areas of reactor performance, including reactor safety, radiation safety, and emergency preparedness. Statistical analysis techniques are applied to each indicator to calculate long-term trends. These trends represent industry averages rather than individual plant performance. No statistically significant adverse trends have been identified in any of the indicators in FY 2011.

5. *Nuclear Material Radiation Exposures*

This measure tracks the number of radiation exposures to the public and occupational workers that exceed Abnormal Occurrence Criterion I.A.3, which is defined as those events that produce unintended permanent functional damage to an organ or a physiological system, as determined by a physician. This measure tracks both nuclear reactors and other nuclear material users, such as hospitals and industrial users. There were no events identified that met the AO Criterion I.A.3 during FY 2011.

6. *Nuclear Material Releases to the Environment*

This measure indicates the effectiveness of the NRC's nuclear material environmental regulatory programs. Exceeding the applicable regulatory limits is defined as a release of radioactive material that causes a total effective radiation dose equivalent to individual members of the public greater than 0.1 roentgen equivalent man (rem) in a year, exclusive of dose contributions from background radiation. No nuclear material releases to the environment that exceeded regulatory limits occurred in FY 2011.

Chairman Jaczko and the Executive Director for Operations R. William Borchardt display the "Best Place to Work in the Federal Government" award

NUCLEAR SAFETY PROGRAMS

The NRC engages in a comprehensive regulatory program that oversees the activities of its licensees. The core of its regulatory program is its licensing and oversight activities. The next sections describe the safety programs the NRC undertook during FY 2011 that resulted in achievement of its Safety goal, strategic outcomes, and performance measure targets for operating reactors, new reactors, fuel facilities, nuclear material users, high-level waste repository, spent fuel storage and transportation, decommissioning and low-level waste, research activities, emergency preparedness and incident response, and international activities.

OPERATING REACTORS

Nuclear Reactor Licensing Activity

The agency's nuclear reactor licensing activity ensures that civilian nuclear power reactors and test and research reactors are operated in a manner that adequately protects public health and safety and the environment while safeguarding radioactive material used in nuclear reactors.

The NRC completed 849 reactor licensing actions in FY 2011. The number of completed licensing action submittals has declined since 2007 because of a significant decrease in the number of licensing actions submitted to the agency. The main reason for the decrease in licensing action submittals by licensees is a result of security enhancements in response to the terrorist attacks of September 11, 2001. These enhancements required an increase in licensing action submittals by licensees. The number of licensing actions has declined as a result of the enhancements. The agency does not expect licensing action submittals to return to the FYs 2001-2007 levels.

During FY 2011, the NRC completed 90.3 percent of the licensing actions in the agency's inventory within one year of receipt and 99.6 percent within two years. The NRC recently began an extensive inspection and licensing effort associated with the reactivation of the Tennessee Valley Authority Watts Bar Unit 2 nuclear power plant. The agency issued a construction permit for this unit in 1973; however, construction was suspended in 1985. Watts Bar Unit 1 received a full power operating license in early 1996, and is presently the last power reactor to be licensed in the U.S. The Tennessee Valley Authority (TVA) suspended construction of Watts Bar Unit 2 in 1985. In August 2007, TVA informed NRC of its plan to resume construction of Watts Bar Unit 2. In FY 2011, the NRC continued its review of the operating license application, which TVA updated in March 2009, and assigned dedicated resident inspectors to monitor TVA's construction activities. The NRC is continuing its reviews of safety, environmental, physical security, and emergency preparedness. The current schedule calls for the NRC to complete its review efforts in 2012.

Power Uprates

The NRC also evaluates nuclear reactor power uprate applications, which allow licensees to safely increase the power output of their plants. The NRC review focuses on the potential impacts of the proposed power uprate on overall plant safety and confirms that plant operation at the increased power level is safe. During FY 2011, the NRC completed four power uprate licensing actions which increased the Nation's electric generating capacity by approximately 211 megawatts, but did not meet its timeliness goals for these reviews in order to resolve key technical issues. The cumulative additional electric power from all power uprates approved since 1977 is approximately 6,021 megawatts. The NRC currently has 16 power uprates under review. If approved, approximately 1,585 megawatts of electric power will be added to the Nation's electric generating capacity.

Chapter 2
PROGRAM PERFORMANCE

The agency expects to receive 27 new power uprate applications in the next five years. If approved, these uprates will add another 1,594 megawatts of electric power to the grid.

License Renewal

The NRC grants reactor operating licenses for 40 years, which can be renewed for additional 20-year periods. The review process for renewal applications is designed to assess whether a reactor can continue to be operated safely during the extended period. To renew a license, the utility must demonstrate that aging will not adversely affect passive, long-lived structures or components important to safety during the renewal period. Additionally, the agency assesses the potential impacts of the extended period of operation on the environment. Inspectors travel to the nuclear power facility to verify the information in the NRC renewal application and confirm that aging management programs have been or are ready to be implemented. Following the safety review, the NRC prepares and makes available to the public a safety evaluation report.

Figure 8
LICENSE RENEWAL APPLICATIONS

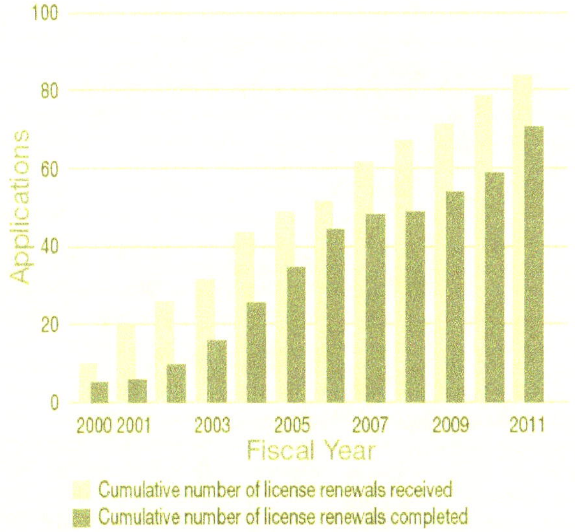

Cumulative number of license renewals received
Cumulative number of license renewals completed

The NRC has received applications to renew the licenses for 84 units at 50 sites since the license renewal program began in 2000 (see Figure 8). It has renewed licenses for 71 units at 41 sites during that time. The NRC is currently reviewing applications to renew the licenses for 13 units at 9 sites. The agency expects that all licensees of currently licensed units will eventually apply to renew their licenses.

Nuclear Reactor Inspection

The NRC provides continuous oversight of nuclear reactors through the Reactor Oversight Program (ROP) to verify that nuclear plants are operated safely and in accordance with the agency's rules and regulations. The NRC performs a rigorous program of inspections at each plant and may perform supplemental inspections and take additional actions to ensure that the plants address significant safety issues. The NRC has at least two full-time resident inspectors at each nuclear power plant site to ensure that facilities are meeting NRC regulations. Inspectors from NRC regional offices and headquarters are also utilized in our inspection program. The NRC has full authority to take action to protect public health and safety, up to and including shutting the plant down. The NRC also conducts public meetings with licensees to discuss the results of the agency's assessments of their safety performance.

The NRC evaluates both inspection findings and performance indicators to assess the performance of each operating nuclear power plant. In FY 2011, more than 99 percent of plant performance indicators were rated green, which is the highest safety rating. In addition, the industry trend indicators for nuclear plants as a whole showed no adverse trends. The results of NRC inspection findings for each plant are documented in inspection reports and are available on the NRC Website http://www.nrc.gov/NRR/OVERSIGHT/ASSESS/pi_summary.html.

In FY 2011, all of the Nation's nuclear power plants operated safely. However, two plants entered the Multiple/Repetitive Degraded Cornerstone Column of the Action Matrix: Browns Ferry Unit 1 and Fort Calhoun.

Browns Ferry Unit 1 transitioned to the Multiple/Repetitive Degraded Cornerstone Column of the Action Matrix in the 4th quarter of CY 2010 (1st quarter of FY 2011) because of one red finding involving the failure to establish adequate testing programs to ensure that motor-operated valves remain capable of performing their safety functions. Because their testing program was inadequate, the licensee failed to detect a valve failure that rendered Loop II of the low pressure coolant injection system incapable of fulfilling its safety function.

Fort Calhoun transitioned to the Multiple/Repetitive Degraded Cornerstone Column of the Action Matrix in the 4th quarter of CY 2010 because of a degraded cornerstone for greater than four quarters associated with a yellow finding originating in the 2nd quarter of CY 2010 and an additional white finding originating in the 2nd quarter of CY 2011. The yellow finding is related to licensee failure to establish and maintain adequate procedures to protect the auxiliary building and intake structure from a significant flood. The white finding is related to licensee failure to identify the cause and preclude a reactor protection system contractor failure resulting in reduced reliability and redundancy of the reactor protection system.

The NRC assesses its inspection program on a regular basis. Assessments conducted in FY 2011 confirm that the agency's ROP met its goal of conducting an objective, risk-informed, and predictable regulatory process that focuses NRC and licensee resources on aspects of plant performance that have the greatest impact on safe plant operations. More information on reactor inspection is available on the NRC Website http://www.nrc.gov/reactors/operating.html.

The NRC Stations Two Inspectors at Every Nuclear Plant

Rulemaking

During FY 2011, the NRC undertook rulemaking activities to improve protection of public health and safety and the environment and enhance effectiveness. The Commission approved a final rule to enhance emergency preparedness requirements for existing nuclear power plants, for those that might be licensed and built in the future, and for research and test reactors. Among the changes in the rule are limitations on the duties of a plant's onsite emergency responders to ensure they are not overburdened during an emergency event and requirements to incorporate hostile-action-based scenarios in the drills and exercise programs. New requirements for back-up measures for alerting and notification systems are also included in the rule. In addition, the new rule requires nuclear power plants to update their evacuation time estimates after every U.S. Census or when changes in population would increase the estimate by either 25 percent or 30 minutes, whichever is less. The agency also published a proposed and final rule to amend its regulations governing the fitness for duty of nuclear power plant workers. The final rule allows licensees the option to use a different method from the one already prescribed in NRC regulations for determining when certain nuclear power plant workers must be afforded time off from work to ensure that such workers are not impaired due to cumulative fatigue.

Chapter 2
PROGRAM PERFORMANCE

The NRC published a final rule to amend 10 CFR 50.55a to include by reference updated ASME code edition and addenda (2005 and later), including two PWR examination code cases, that provide requirements for the construction, operation, in service inspection, and in service testing of nuclear power plants. Additionally the agency published a proposed rule for public comment that would implement new authority for access to enhanced weapons and associated firearms background checks provided in Section 161A of the AEA.

Investigations and Enforcement

Licensee compliance with NRC requirements plays an important role in ensuring that safety is being maintained. NRC policies deter noncompliance and encourage prompt identification and timely, comprehensive corrective actions. Licensees, contractors, and their employees who do not achieve the high standard of compliance expected by the NRC are subject to enforcement sanctions and investigations of potential willful violations. Each enforcement action depends on the circumstances of the case. The NRC will not permit licensees to continue to conduct licensed activities if they cannot achieve and maintain adequate levels of safety. In FY 2011, the NRC conducted 21 escalated enforcement actions and 160 opened investigations of potential willful wrongdoing.

NUCLEAR POWER PLANT REVIEW

On March 11, 2011, Japan experienced a severe earthquake which resulted in the shutdown of multiple nuclear reactors. This earthquake was followed by a tsunami that inflicted catastrophic damage to the coastline of Japan. At the Fukushima Dai-ichi nuclear site, the earthquake and tsunami caused the loss of all alternating current power. The sustained loss of electrical power led to damage to nuclear fuel and radiological releases off site. Following the accident, the Commission directed NRC staff to conduct a systematic and methodical review of NRC processes and regulations

to determine whether the agency should make additional improvements to its regulatory system, and to provide recommendations to the Commission for its policy direction. The Commission directed that this review include both near-term and longer-term components.

Environmental Minister of Japan, Goshi Hosono, at the NRC Operations Center

The NRC's near-term review was completed on July 12, 2011. The review found that, based on the current NRC regulations and domestic nuclear plant capabilities, a sequence of events like the Fukushima accident is unlikely to occur in the United States. Therefore, continued operation and licensing activities do not pose an imminent risk to public health and safety. Notwithstanding, the near-term review identified 12 recommendations that are intended to clarify and strengthen the regulatory framework and enhance safety through protection against natural disasters, mitigation, and emergency preparedness. The Commission's direction to the staff regarding these recommendations can be found on the NRC Website http://www.nrc.gov/reading-rm/doc-collections/commission/srm/2011/2011-0093srm.pdf. The agency is reviewing strategy recommendations developed in two Commission papers (ML11245A127 and ML11269A204).

NEW REACTORS

The NRC reviews applications for new reactor facilities submitted by prospective licensees and issues standard design certifications, early site permits, limited work authorizations, construction permits, operating licenses, and combined licenses, when appropriate. At present, the NRC anticipates that these activities will involve new light-water reactor (LWR) facilities in a variety of projected locations throughout the United States.

Briefing to Commission on Severe Accidents

New Reactor Design Certification

The NRC is reviewing three design certification (DC) applications and two design certification amendments. By issuing a DC, the NRC approves a nuclear power plant design independent of an application to construct or operate a plant. A DC is valid for 15 years from the date of issuance, but can be renewed for an additional 10 to 15 years.

The NRC continued reviewing DCs for the General Electric Economic Simplified Boiling-Water Reactor design (ESBWR), the AREVA Evolutionary Power Reactor, and Mitsubishi's U.S. Advanced Pressurized-

Water Reactor. The agency has continued the process of performing a DC amendment review for the Westinghouse Advanced Passive (AP) 1000 and a DC amendment review for the Advanced Boiling-Water Reactor (ABWR) DC amendment. The purpose of the AP1000 DC amendment is to replace the combined operating license (COL) information items and design acceptance criteria (DAC) with specific design information, address the effects of the impact of a large commercial aircraft, incorporate design improvements, and increase standardization of the design. The purpose of the ABWR amendment is to address the requirements in 10 CFR 50.150, the Commission's new aircraft impact rule.

The NRC prepared and issued the proposed rulemakings for the AP1000 design certification amendment, the ABWR design certification amendment, and the General Electric ESBWR Reactor DC and is preparing the final rulemaking packages for these rulemakings. In addition, the agency issued the final design approval for the General Electric ESBWR DC application.

Early Site Permits

The NRC approves the site for a nuclear facility by issuing an early site permit. Early site permits are valid for 10 to 20 years and can be renewed for an additional 10 to 20 years. The NRC review of an early site permit application addresses site safety issues, environmental protection issues, and plans for coping with emergencies, independent of the review of a specific nuclear plant design.

In FY 2011, the NRC began its safety and environmental reviews of two early site permit applications that were submitted in FY 2010. Specifically the two early site permits that are in review are the Victoria County Station early site permit application submitted by Exelon Nuclear Texas Holdings, LLC, for a site located in Victoria

County, TX, and by PSEG Power, LLC, and PSEG Nuclear, LLC, for a site adjacent to the Salem and Hope Creek Generating Stations now in operation in Lower Alloways Creek, Salem County, NJ. The NRC initiated pre-licensing activities for the Blue Castle early site permit application expected in FY 2012/13.

Combined Operating License

A COL authorizes construction and operation of a nuclear power plant. The application for a COL must contain essentially the same information required in an application for an operating license, including financial and antitrust information and an assessment of the need for power. The application must also describe the Inspections, Tests, Analyses, and Acceptance Criteria (ITAAC) that are necessary to ensure that the plant has been properly constructed and will operate safely.

Vogtle New Reactor Site

The NRC has two objectives for the review of COL applications. The first objective is to ensure that the proposed new reactor designs and planned operations will be in accordance with NRC regulations for safety, security, and the environment. The second objective is that the reviews will be completed on the schedules negotiated with applicants. To date, the agency has docketed 18 COL applications that have been filed by

the nuclear power industry for sites across the country. Twelve of the eighteen applications are being actively reviewed. In response to applicant requests, the agency has suspended the reviews of the other six applications: Grand Gulf, Victoria County Station, Callaway, Nine Mile Point, River Bend, and Bellefonte Units 3 & 4. One of these applicants submitted an early site permit application for a site located in Victoria County, TX, as noted above, and requested that the Victoria County Station COL be withdrawn after the acceptance of the early site permit application. The agency did not receive any new COL applications in FY 2011.

In FY 2011, the NRC completed the environmental review of four COL combined license applications through the issuance of the Final Environmental Impact Statements for the following COL applications: South Texas Project, V.C. Summer, Calvert Cliffs, and Comanche Peak. In addition, the staff completed the final supplemental environmental impact statement for the Vogtle COL application. The Commission conducted its first Mandatory Hearing for the Vogtle COL application in September, 2011. The second such hearing was held in October, 2011 to review the COL application for the V.C. Summer project.

In FY 2011, the NRC continued to enhance the regulatory framework for COLs to clarify requirements for licensees. The NRC issued the following five interim staff guidance (ISG) documents for COLs: (1) ISG 18: "Reliability Assurance Program, Section 17.4 of the Standard Review Plan," (issued final); (2) ISG 21: "Review of Nuclear Power Plant Designs Using a Gas Turbine Driven Standby Emergency Alternating Current Power System," (issued final); (3) ISG 22: "Impacts of Changes During Construction," (issued for comments, staff anticipates issuance as final in early FY 2012); (4) ISG 19: "Review of Evaluation to Address Gas Accumulation Issues in Safety Related Systems," (issued final); and

(5) ISG-25: "Changes During Construction," (issued for comments, staff anticipates issuance as final in early FY 2012).

Oversight

The NRC has in place the structure and procedures required to conduct new reactor construction oversight for plants to be licensed under 10 CFR Part 52, "Licenses, Certifications, and Approvals for Nuclear Power Plants," and has begun executing construction inspection activities for Vogtle Units 3 & 4. The process for oversight of new reactor construction has been documented in Inspection Manual Chapters and inspection procedures. All inspection procedures that are required to implement inspections of licensee activities related to ITAAC have been approved and issued for use. The agency continued to make significant progress in the development and improvement of programs and procedures to support inspection of activities occurring later in construction. For example, the agency has refined the ITAAC closure process to include the maintenance of closed ITAAC. The agency also continued development of: (1) inspector training, (2) business processes to support information technology system needs, (3) generic inspection schedules, and (4) enhancements to the existing assessment and enforcement program for new reactors. In addition, the NRC maintained an aggressive schedule of public meetings to provide a forum for stakeholders to participate and comment on staff proposals for ITAAC closure, licensee assessment, enforcement, and other construction inspection program topics.

The NRC has placed a construction senior resident inspector and resident inspector at the Vogtle 3 & 4 construction site and has conducted multiple inspections of the quality assurance program associated with the limited work authorization activities. In FY 2011, the NRC completed the first semiannual performance review of Vogtle Units 3 and 4, which covered the period between July 1, 2010, and December 31, 2010. The NRC transitioned to an annual performance review cycle effective beginning January 1, 2011. The NRC completed a midcycle review which covered the time period between January 1, 2011, and June 30, 2011. Plant performance for Vogtle Units 3 and 4 for both review periods was within the Baseline Program column of the NRC's Construction Action Matrix, based on all inspection findings being categorized at Severity Level IV or lower.

The NRC maintains a regular schedule of vendor inspections and an active program of international cooperation to support increased fabrication activities domestically and internationally in response to new reactor construction plans. The agency conducts these inspections to ensure the effective implementation of quality assurance program requirements imposed on vendors by NRC applicants and licensees. In FY 2011, the NRC completed 18 inspections.

Additionally, international cooperative efforts have included technical discussions with foreign regulatory counterparts, sharing vendor experience and other information with other countries, NRC inspector rotations to facilities under construction in other countries, and participation in the Multinational Design Evaluation Program (MDEP) and the Nuclear Energy Agency Committee on Nuclear Regulatory Activities Working Group on the Regulation of New Reactors. Exchanges such as these have provided key insights into each country's methods of oversight and have enabled the agency to build a foundation of trust and a rapport for communicating and sharing key information and findings.

Chapter 2
PROGRAM PERFORMANCE

Advanced Reactor Program

During FY 2011, the NRC continued its efforts to support congressionally-mandated and the DOE sponsored programs such as the Next Generation Nuclear Plant, while also supporting efforts related to the growing commercial interest in integral pressurized-water reactors. These efforts included developing the regulatory framework necessary to support the review of applications for these new reactors. The NRC has continued its strong outreach to conduct pre-application interactions with stakeholders and potential applicants.

The agency has continued to focus on the identification and resolution of generic policy issues as well as key technical issues for the licensing of small modular reactor (SMR) designs while concurrently training its staff to be prepared for the review of potential future SMR applications. Issues considered have included emergency preparedness and security considerations in view of the recent events in Japan and lessons learned. The agency has met with the first utility planning to construct SMRs and published several papers outlining plans to resolve some of the policy and technical issues. In addition, the agency staff promulgated a regulatory issue summary asking for voluntary responses from companies interested in submitting applications for SMRs to help effectively plan resources.

FUEL FACILITIES

Licensing

The NRC licenses and inspects all commercial nuclear fuel facilities that process and fabricate uranium concentrates into the reactor fuel which powers the Nation's nuclear reactors. Licensing activities include detailed health, safety, safeguards, and environmental licensing reviews of licensee programs, procedures, operations, and facilities to ensure safe and secure operations.

The agency continued its safety, security, and environmental reviews of two license applications for uranium enrichment facilities in FY 2011. These facilities increase the concentration of the uranium 235 isotope from its natural enrichment of about 0.7 percent of natural uranium to four to five percent. The uranium is used in commercial power reactors, such as those used throughout the commercial power industry in the United States. The first application, submitted in December 2008 by AREVA, is for a centrifuge enrichment facility to be built near Idaho Falls, ID. The second, submitted in June 2009 by General Electric-Hitachi, is for a laser-based enrichment facility to be built in Wilmington, NC.

The agency completed the review for the AREVA Enrichment Services Eagle Rock license application for the Eagle Rock Enrichment Facility. The agency found the record sufficient and the staff review adequate to support 10 CFR 30, 40 and 70 findings for license approval. The evidentiary hearing for the Final Environmental Impact Statement was held in July 2011 in Idaho Falls.

In December, 2010, the agency approved the issuance of the final Safety Evaluation Report for the license application by Shaw AREVA MOX Services, LLC, to possess and use radioactive material at the Mixed-Oxide Fuel Fabrication Facility at the DOE's Savannah River Site near Aiken, SC. The Safety Evaluation Report was reviewed and approved.

A byproduct of uranium enrichment is depleted (i.e., reduced in the uranium 235 isotope) uranium hexafluoride. During FY 2011, the agency accepted a license application to construct and operate a facility to convert depleted uranium hexafluoride into an oxide form for ultimate disposal and to recover the fluorine for other commercial applications.

Oversight

The NRC's fuel cycle oversight process consists of both planned and reactive inspections with enforcement and periodic assessments based on the findings of these inspections. The agency has full authority to take action to protect public health and safety, up to and including, shutting down the facility.

Construction at the URENCO Enrichment Facility

The NRC conducted a thorough review of the root and contributing causes of events that occurred in October 2009 at the Nuclear Fuel Services facility in Erwin, TN. The event did not cause a release of hazardous material to the environment and had no public health and safety consequences. The agency's review of the underlying causes and the licensee's response to the events led to a licensee commitment to shutdown process lines and to keep them shut down until the agency approved their restart. The agency confirmed the commitment by issuing a confirmatory action letter. This confirmatory action letter established the corrective measures to be taken by the licensee before seeking agency approval to restart process lines. As the licensee identified its readiness to restart each process line, the agency conducted additional inspections to verify its readiness for restart and supplemental inspections during the initial

operation of each process line as it was restored to service. The fifth and final process line was authorized for restart in July 2011.

Rulemaking

In response to sustained industry interest in reprocessing spent nuclear fuel, the NRC continued to work on developing a technical basis for rulemaking to establish the regulatory framework for licensing a reprocessing facility. In FY 2009, the agency completed a review to identify and prioritize gaps in the existing regulations. During FY 2011, the agency continued to define the technical basis needed to support the development of proposed regulations to resolve the identified gaps and establish an effective and efficient regulatory framework.

The NRC continues to conduct rulemakings to secure special nuclear material. In FY 2009, the agency began an initiative to revise and consolidate the regulations for material control and accounting of special nuclear material. During FY 2010, staff started developing the draft rule text. This work continued in FY 2011, and the draft text is expected to be released for public comment in FY 2012.

A proposed rule and draft guidance to require an Integrated Safety Analysis (ISA) for certain Part 40 facilities was published in the Federal Register on May 17, 2011. The public comment period ended in September 2011 and the staff is considering and resolving public comments. The staff expects to finalize the rule in mid-2012

Investigation and Enforcement

The NRC will not permit licensees to conduct licensed activities if they cannot achieve and maintain adequate levels of safety and security. The agency assesses compliance, undertakes enforcement actions, and investigates potential willful violations. For fuel facilities, the agency conducted eight open investigations of potential willful wrong-doing and five escalated enforcement actions in FY 2011.

NUCLEAR MATERIALS USERS

The NRC licenses and inspects the commercial use of nuclear material for industrial, medical, and academic purposes. Commercial uses of nuclear materials include medical diagnosis and therapy, medical and biological research, academic training and research, industrial gauging and nondestructive testing, production of radiopharmaceuticals, and fabrication of commercial products (such as smoke detectors) and other radioactive sealed sources and devices. The agency currently regulates about 3,000 specific licensees for the use of radioactive materials. Under NRC's Agreement State program, 37 States assumed regulatory responsibility over approximately 19,200 licenses for the industrial, medical, and other users of nuclear materials in their States. The agency reviews the Agreement State programs as well as certain NRC licensing and inspection programs through the Integrated Materials Performance Evaluation Program.

Detailed health and safety reviews of license applications, as well as inspections of licensee procedures, operations, and facilities, provide reasonable assurance of safe operations and the production of safe products. The NRC routinely inspects nuclear material licensees to ensure

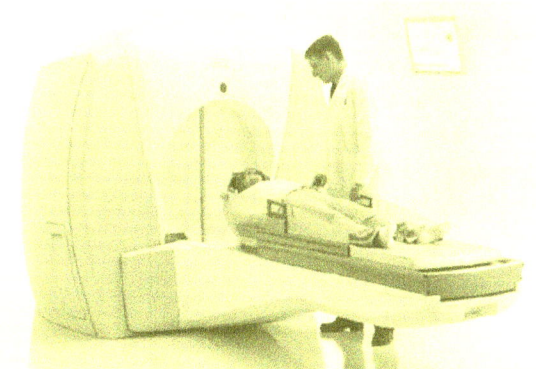

Gamma Knife Used in Medical Procedures

that they are using nuclear materials safely, maintaining accountability of those materials, and protecting public health and safety. The agency also analyzes operational experience from NRC and Agreement State licensees and regularly evaluates the safety significance of events reported by licensees and Agreement States.

Licensing and Oversight

The NRC completed 2,104 materials licensing actions and 1,010 routine health and safety inspections in FY 2011. The agency maintained its high standards with timely reviews of nuclear material license renewals and sealed-source and device designs in FY 2011. The agency completed 96 percent of new application and license amendment reviews within 90 days of receipt and 97 percent of license renewal and sealed-source and device design reviews within 180 days of receipt.

Rulemaking

The NRC proposed to amend its regulations that govern the licensing and distribution of byproduct materials aimed at making regulations clearer, more risk-informed, and up-to-date in FY 2011. An agency working group began to prepare a proposed rule for public comment, expected to be published in the *Federal Register* in FY 2013. The rule proposes the following changes in 10 CFR Part 35: modifying preceptor attestation requirements; extending grandfathering to certified individuals that were named in Part 35 prior to October 25, 2005; naming associate or assistant Radiation Safety Officers on an NRC medical-use license; and a likely change in the definition of a medical event including revised reporting and notifications of medical events for permanent implant brachytherapy. The agency conducted a special series of facilitated public workshops to engage stakeholders on possible revisions of the agency's radiation protection requirements in light of international recommendations.

Investigation and Enforcement

The NRC will not permit licensees to conduct licensed activities if they cannot achieve and maintain adequate levels of safety. For nuclear materials users, the agency conducted 34 opened investigations of potential willful wrong-doing and 40 escalated enforcement actions in FY 2011.

HIGH-LEVEL WASTE REPOSITORY

Prior to FY 2011, NRC staff had been conducting a review and formal hearing on a license application from the DOE to dispose of high-level waste underground in a deep geologic repository. In 2010, DOE filed a motion with the NRC Atomic Safety and Licensing Board seeking to withdraw its license application. Subsequently, the NRC Licensing Board denied DOE's motion. In FY 2011, the NRC completed documentation and knowledge management activities to preserve license application review material and lessons learned from the Yucca Mountain license application review. The NRC Licensing Board took steps to shutdown the Las Vegas, Nevada Hearing Facility and the licensing support network, which makes documentary material available electronically to parties and interested governmental participants to the hearing on the license application. The Licensing Board has taken reasonable measures to ensure that documents necessary for the proceeding are maintained in a format that is easily acceptable for all parties. In September 2011, the Commission affirmed an order stating that they were evenly split on whether to take the affirmative action of overturning or upholding the licensing board's decision to deny DOE's motion to withdraw the license application. In the order, the Commission directed the licensing board to complete all necessary and appropriate case management activities, including disposal of all matters currently pending before it and comprehensively documenting the full history of the adjudicatory proceeding. The licensing board suspended the licensing hearing in an order issued on September 30, 2011.

SPENT FUEL STORAGE AND TRANSPORTATION

The NRC ensures that spent nuclear fuel is safely stored and transported. The agency conducts licensing and certification reviews to ensure that spent fuel storage facility and cask designs and domestic and international shipments of spent fuel and other risk-significant radioactive materials are safe and secure and comply with agency regulations.

Shipments of radioactive materials are safely and securely transported each year within the U.S. Several Federal agencies share responsibility for regulating the safety and security of those shipments. The NRC closely coordinates its transportation-related activities with those of the U.S. Department of Transportation and, as appropriate, DOE. The agency inspects vendors, fabricators, and licensees using transport packages, spent fuel storage casks, and interim storage of spent fuel both at and away from reactor sites to help ensure the safety and security of spent fuel storage and transportation.

Licensing and Oversight

In FY 2011, the NRC completed 50 transport package designs and nine storage cask and facility design reviews. The review of transportation and interim storage licensing requests ensures that shipments are made in NRC-approved packages that meet rigorous performance requirements and verifies that spent fuel is safely stored, thereby enabling continued reactor and decommissioning operations. The agency also conducted 19 inspections of activities related to radioactive material package certificate holders, spent fuel storage cask certificate holders, and inspections at independent spent fuel storage facilities to ensure that casks are being designed, fabricated, and used according to approved safety requirements.

Spent Fuel Dry Cask Storage

Rulemaking

In FY 2011, the NRC published changes to its regulations concerning licensing requirements for the independent storage of spent nuclear fuel, high-level radioactive waste, and reactor-related greater than Class C waste. The amendments extend and clarify the license terms for dry storage cask certificates of compliance (CoCs) and independent spent fuel storage installation (ISFSI) licenses. The amendments also require certain aging management requirements for both specific license and CoC renewals. Finally, the amendments allow general licensees under 10 CFR Part 72, "Licensing Requirements for the Independent Storage of Spent Nuclear Fuel, High-Level Radioactive Waste, and Reactor-Related Greater than Class C Waste," to implement changes authorized by a later CoC amendment to a cask loaded under the initial CoC or an earlier CoC amendment. The rule changes improve the regulatory efficiency of 10 CFR Part 72.

The NRC developed a plan for integrating spent nuclear fuel regulatory activities to more effectively address the regulatory and licensing aspects of extended storage and transportation, reprocessing, and disposal of spent nuclear fuel and high-level waste. The purpose of the plan is to ensure that regulation of the back end of the fuel cycle accomplishes safety, security, and environmental protection in an efficient and effective manner and that decisions made about one component or area of this system adequately consider other components or areas (i.e., treating spent fuel and high-level waste regulation as a system of interrelated activities). By integrating the approach for regulation of spent nuclear fuel or high-level waste, the agency can improve the efficiency and effectiveness of its regulatory processes and gives stakeholders stability and predictability in a dynamic environment.

In FY 2011, the Commission approved publication of proposed rule changes for public comment that would establish security requirements for the physical protection of irradiated reactor fuel in transit. The proposed rule would establish the acceptable performance standards for the protection of spent nuclear fuel from theft, diversion, or radiological sabotage, and would replace orders previously imposed by the Commission after September 11, 2001.

The NRC also began a comprehensive review of the spent fuel storage and transportation regulatory programs to evaluate their adequacy for ensuring safe and secure storage of spent fuel for extended periods beyond 120 years, including research to enhance the regulatory framework in support of extended periods.

DECOMMISSIONING AND LOW-LEVEL WASTE

Decommissioning removes radioactive contamination from buildings, equipment, ground water, and soil, achieving levels that permit the release of the property while protecting the public. The NRC terminates the licenses for decommissioned facilities after the licensees demonstrate that the residual onsite radioactivity is within regulatory limits and sufficiently low to protect the health and safety of the public and the environment. Completion of decommissioning, environmental, and performance assessment activities ensures that residual radioactivity does not pose an unacceptable risk to the public.

Decommissioning

The NRC has completed decommissioning at 19 materials sites and nine power or research reactors for a total of 28 sites since 2006. In FY 2011, the agency oversaw decommissioning activities at approximately 85 power and early demonstration reactors, research and test reactors, uranium recovery sites, complex materials sites, and fuel cycle facilities. Additionally, the NRC published a final rule amending its regulations to improve decommissioning planning and thereby reducing the likelihood that a current operating facility will become a legacy site. The agency increased its activities at military sites containing Naturally Occurring Radioactive Material and Army sites with depleted uranium contamination from military munitions. The agency continued its emphasis on the decommissioning of legacy uranium recovery sites during FY 2011 and began several initiatives to improve the program, including updating guidance and enhancing communication with DOE, States, Native American Tribes, and stakeholders.

Uranium Recovery Licensing and Oversight

The NRC conducts regulatory oversight at eight operational uranium recovery sites and reviews and approves, if regulations are met, the applications for new, restarting, or expanding uranium recovery facilities. In FY 2011, the agency had eight applications for new, restarts, or expanding uranium recovery facilities in-house. The agency worked on five of those applications in FY 2011. These reviews included both safety and environmental reviews. The agency published the final supplemental environmental impact statements, published the final safety evaluation reports, and approved the applications for both the Nichols Ranch and Lost Creek uranium recovery facilities in FY 2011.

Low-Level Waste

The NRC conducts regulatory activities to help ensure the safe management and disposal of low-level radioactive waste generated by radioactive material users, nuclear power plants, and other NRC licensees. The agency performed monitoring visits and issued reports for the DOE's Savannah River Site Saltstone facility and the Idaho National Laboratory. In addition, the agency also conducted outreach with stakeholders and licensees on issues related to issuing guidance on how to classify waste for disposal and potential draft rule language for a proposed change to 10 CFR Part 61 for site evaluation prior to receiving either long-lived or blended wastes.

RESEARCH ACTIVITIES

The NRC's safety research program evaluates and resolves safety issues for nuclear power plants and other facilities and materials that the agency regulates. The agency conducts its research program to evaluate existing and potential safety issues; supply independent expertise, information, and technical judgments to support timely and realistic regulatory decisions; reduce uncertainties in risk assessments; and develop technical regulations

THE IN SITU URANIUM RECOVERY PROCESS

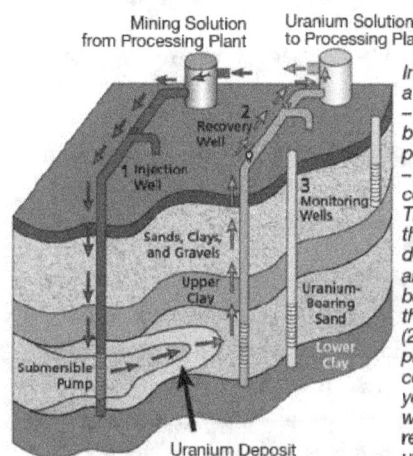

Injection wells (1) pump a chemical solution – typically sodium bicarbonate, hydrogen peroxide, and oxygen – into the layer of earth containing uranium ore. The solution dissolves the uranium from the deposit in the ground and is then pumped back to the surface through recovery wells (2) and sent to the processing plant to be converted into uranium yellowcake. Monitoring wells (3) are checked regularly to ensure that uranium and chemicals are not escaping from the drilling area.

and standards. When possible, the agency engages in cooperative research with other government agencies, the nuclear industry, universities, and international partners.

In FY 2011, the NRC research program addressed key areas that support the agency's safety mission. Some of the more important issues addressed include: verification and validation of fire safety models; evaluation of material degradation of reactor system and pressure boundary components, especially as it relates to license renewal periods; material degradation research on decommissioning facilities focused on long-term performance of concrete and soil materials used as barriers; evaluation of digital systems to analyze failure modes; research on hazards from natural events, including seismic hazard issues, flooding, and tsunami events; advanced reactor research; development of advanced tools for probabilistic risk assessment activities that support risk-informed regulatory decision-making; and severe reactor accident consequence analyses.

Fire Safety

The NRC has continued conducting collaborative research to develop state-of-the-art knowledge, guidance, methods, and tools in support of regulatory activities related to fire protection and fire risk analyses. This collaborative research included participation from the Electric Power Research Institute, the National Institute of Standards and Technology, Sandia and Brookhaven National Laboratories, and the University of Maryland. The NRC and the Electric Power Research Institute continue to provide training on NUREG/CR 6850, "EPRI/NRC RES Fire PRA Methodology for Nuclear Power Facilities," issued in September 2005, for those nuclear power plants that have submitted letters of intent to transition to National Fire Protection Association Standard 805, "Performance-Based Standard for Fire

Protection for Light Water Reactor Electric Generating Plants," via 10 CFR 50.48(c). This research has been the basis for NRC's moving forward on risk-informed, performance-based fire protection requirements for facilities regulated by the agency.

Cable Fire Test

Reactor Safety Code Development

The NRC uses computer codes to perform probabilistic risk assessments and evaluate thermal-hydraulic conditions, severe accidents, fuel behavior, and reactor kinetics during various operating and postulated accident conditions. Results from such analyses support decision-making for risk-informed activities, the review of licensees' codes and performance of audit calculations, and the resolution of other technical issues. Code development is directed toward improving the realism and reliability of code results and making the codes easier to use.

Advanced Reactor Research

In response to the *Energy Policy Act of 2005* (EPA), the NRC initiated research in a number of major technical areas related to licensing a prototype high-temperature gas-cooled reactor (HTGR), otherwise referred to as the Next Generation Nuclear Plant (NGNP), which can be used to generate electricity, hydrogen, and/or process heat for industrial applications. The agency published the "High Temperature Gas-Cooled Reactor Research Plan." Some work, which had been initiated upon the enactment of the EPA, has been completed, including development of HTGR preliminary plant models for incorporation into NRC's safety analysis code, scoping analysis of important HTGR thermal-fluids phenomena using system codes as well as computational fluid dynamics tools, and modification of LWR-specific reactor physics codes for HTGR nuclear analysis applications. The NRC initiated development of an evaluation model for safety evaluation of HTGR, and a preliminary fuel performance model. The agency also initiated research in other major technical areas pertaining to HTGR, notably graphite and high-temperature metallic materials, structural integrity assessment, instrumentation and control, human factors engineering, and probabilistic risk assessment.

The NRC has also begun to generate models for its thermal-hydraulic and severe accident codes to support review of the new integral pressurized-water reactor (iPWR) designs. The agency developed generic iPWR models that can be used to explore postulated event sequences to support pre-application activities.

Materials Degradation

The NRC continues to research materials degradation issues for currently licensed reactors and waste and decommissioning facilities. The purpose of this research is to identify susceptible materials and assess component-specific degradation mechanisms in existing reactors and waste and decommissioning facilities to ensure continued safe operation. The agency is also performing research on reactor internals to determine the effects of neutron fluence and thermal effects on the physical properties of reactor internal materials. The long-term performance of concrete and soil materials that are used to contain or restrict the movement of radioactive contaminants has been the research focus for decommissioning facilities. Cooperative work with the DOE and National Institute of Standards and Technology has been particularly effective in improving the understanding of degradation mechanisms in concrete, and work with the U.S. Geological Survey has proved invaluable in addressing degradation in covers on disposal sites. In addition, the agency is conducting research into potential technical issues that may challenge long-term safe operation of existing commercial nuclear power plants in second and subsequent license renewal periods.

Digital Instrumentation and Control

The NRC's research supports the licensing of new digital instrumentation and control systems intended for use in retrofits to operating reactors and for use in new and next-generation reactors. The agency is actively engaged in ongoing research involving identifying and analyzing digital system failure modes. In FY 2011, the agency published Research Information Letter-1001, "Software-Related Uncertainties in the Assurance of Digital Safety Systems – Expert Clinic Findings, Part 1," which supports the judgment exercised in licensing reviews of complex digital safety systems. NRC also published NUREG/IA-0254, "Suitability of Fault Modes and Effects Analysis (FMEA) for Regulatory Assurance of Complex Logic in Digital Instrumentation and Control Systems," which investigates the use of FMEA analysis of software in licensing reviews.

Chapter 2
PROGRAM PERFORMANCE

Probabilistic Risk Assessment

The NRC continues to research the development of advanced models, methods, and tools for probabilistic risk assessment activities that support risk-informed regulatory decision-making. The Standardized Plant Analysis Risk (SPAR) models and the Systems Analysis Program for Hands-on Integrated Reliability Evaluations (SAPHIRE) computer code support the agency's risk-informed programs such as the Accident Sequence Precursor Program, Incident Investigation Program, and the Significance Determination Process. In FY 2011, the agency continued to improve the capabilities and usability of the SAPHIRE software that allows analysts to perform probabilistic risk assessments for nuclear power plants and other complex systems, facilities, or processes. The agency is strengthening the technical basis for the SPAR models and expanding the model scope to include additional hazard categories such as fires, floods, and seismic events. The agency is also investigating methods to incorporate new digital instrumentation and control systems (hardware and software) into nuclear power plant risk assessments.

Natural Hazards Research

The NRC is researching seismic hazard issues to support the siting of new reactors and the evaluation of the seismic safety of existing nuclear facilities. In cooperation with academic institutions, other Federal and State agencies, and industry, the NRC is conducting a program to develop ground motion propagation and earthquake source zone models.

The NRC is also conducting a study of potential tsunami sources and the resulting potential hazards to NRC-regulated facilities in collaboration with the U.S. Geological Survey and the National Oceanographic and Atmospheric Administration. The agency is using the results of this research to inform licensing decisions and update risk assessments.

The agency is also conducting research on flooding events, including estimating the severity of natural events such as coastal storm surge from hurricanes, local inland flooding from extreme precipitation events or combinations of precipitation, dam break, and/or seasonal snow melt. The NRC is working with the U.S. Army Corps of Engineers and the Department of Interior's Bureau of Reclamation to update databases and guidance documents that are more than 30 years old to support the use of the latest analytical techniques. The Corps of Engineers is focused on the estimation of storm surge for the Gulf of Mexico and South Florida, which complements and benefits from initiatives in their own agency's programs. The Bureau of Reclamation is updating information for selected areas of the country covered by the National Weather Service's Hydrometeorological Report 51 (HMR 51) for maximum precipitation events in the eastern U.S. where most new plants are planned.

State-of-the-Art Reactor Consequence Analysis

The NRC continues to develop a body of knowledge on the realistic outcomes of severe reactor accidents for two pilot plants, Peach Bottom and Surry, under the State-of-the-Art Reactor Consequence Analysis (SOARCA) project. In addition to incorporating state-of-the-art modeling, one objective of the SOARCA study is to incorporate plant improvements not reflected in earlier assessments (e.g., hardware, procedures, security-related enhancements, emergency planning) as well as plant updates in the form of power uprates and higher core burnup. NRC is currently addressing comments from an independent peer review panel of subject matter experts on a technical evaluation of two commercial reactors completed last fiscal year, and expects to issue the draft report for public review and comment in the second quarter of FY 2012. In FY 2011, the agency briefed the Organization of Agreement States on the SOARCA project, and presented the project at the 2011 Regulatory Information Conference.

NRC EMERGENCY PREPAREDNESS AND INCIDENT RESPONSE

The NRC's emergency preparedness and incident response activities ensure that adequate measures can and will be taken to mitigate plant events, to minimize possible radiation doses to members of the public, and to ensure that the agency can respond effectively to events at licensee sites.

In FY 2011, the NRC supported the U.S. Government's response to the events at Japan's Fukushima Dai-ichi nuclear facility and coordinated its actions with other Federal agencies as part of the U.S. Government's response. NRC emergency responders staffed the Headquarters Operations Center (HOC) for over three months and closely monitored the status of the Fukushima Dai-ichi reactors and spent fuel pools. The extreme circumstances at the plant led to a fast-paced response effort with a large degree of uncertainty about plant conditions. The agency was a key contributor to U.S. Government efforts aimed at ensuring that the U.S. citizens living in the region were safe. In responding to this unique challenge, the agency identified a number of good practices and lessons learned items that will be used to improve the NRC's response program. The event in Japan also demonstrated the ability of the HOC's Information Technology systems to support continuous and effective response operations.

The NRC also participated in several exercises in FY 2011. The agency participated in the National Level Exercise (NLE 11), the annual continuity exercise (Eagle Horizon 11) for Federal Executive Branch departments and agencies, and in the several security-related tabletop exercises focused on the Safeguards Team. The NRC also hosted a multiagency senior official tabletop exercise that focused on the challenges of recovering from the events related to a reactor-accident scenario at a nuclear power plant.

In August 2011, the Commission approved a final rule that enhances the emergency preparedness regulations. The enhancements to the regulations include codifying voluntary industry efforts since September 11, 2001. The proposed final Emergency Preparedness rule was made publicly available in April, 2011.

Consistent with its policy to provide States with potassium iodide as requested, the NRC worked with States to replenish potassium iodide supplies to be used as a supplement to public protective actions within the 10-mile emergency planning zones around nuclear power plants.

Emergency Response Drill in Region III

The NRC continued its modernization of the Emergency Response Data System, which provides real-time information from nuclear power plants to the NRC and State operations centers during emergencies. The modernization of this system enhances cyber security and reliability and includes improvements to the user interface.

INTERNATIONAL ACTIVITIES

The NRC's international responsibilities include participation in activities that support U.S. Government compliance with international treaties and agreements; export and import licensing of nuclear facilities, equipment, and materials; programs of bilateral nuclear cooperation and assistance; and multinational nuclear safety organizations such as the International Atomic Energy Agency (IAEA) and the Organization for Economic Co-operation and Development's Nuclear Energy Agency (NEA). The agency is also the U.S. representative to the IAEA's radiation, waste, transportation and nuclear safety standards committees and NEA's technical standing committees.

International Treaties and Agreements

In April 2011, the NRC led the U.S. delegation in the Review Meeting of Contracting Parties to the Convention on Nuclear Safety, including participating in an extraordinary meeting to discuss the effects of the March 2011 earthquake and tsunami in Japan on the Fukushima Nuclear Power Plant. The NRC is participating in U.S. Government activities to prepare for the 2012 Review Meeting of Contracting Parties to the Joint Convention on the Safety of Spent Fuel Management and the Safety of Radioactive Waste Management.

Export and Import Licensing

The NRC issued a final rule updating 10 CFR Part 110, "Export and Import of Nuclear Equipment and Material," to reflect obligations to the IAEA and the recent IAEA publication of INFCIRC/225/Revision 5, "Nuclear Security Recommendations on Physical Protection of Nuclear Material and Nuclear Facilities."

The NRC completed reviews for, and issued as appropriate, 139 import/export licensing actions, seven reviews of Executive Branch proposed subsequent arrangements, and 19 reviews of Executive Branch proposed Part 810 approvals. NRC participated in six U.S. interagency bilateral physical protection visits to support export licensing. The NRC's import/export licensing reviews ensure that nuclear equipment and material are transported and used in a manner consistent with applicable U.S. law and international requirements.

Bilateral Cooperation and Assistance

In FY 2011, new Arrangements of Cooperation and Assistance were signed with The American Institute in Taiwan and The Taipei Economic and Cultural Representative Office (AIT-TECRO) - Taiwan and existing Arrangements were renewed with the regulatory bodies of Romania and Slovenia. In addition, on March 8, 2011, a Memorandum of Cooperation was signed with the Korea Institute of Nuclear Safety (KINS), during the Nuclear Regulatory Commission's Regulatory Information Conference (RIC). This is the first Memorandum of Cooperation of this type, wherein NRC and KINS have defined areas of cooperation and assistance to focus on establishing specific outreach initiatives for countries interested in new nuclear power programs. On July 3, 2011, the Arrangement between NRC and the Israeli Atomic Energy Commission (IAEC) was signed by both the NRC and the IAEC. The agency continues an active program for bilateral cooperation and assistance. For example, the agency continued cooperation with China on the regulatory aspects for the first-of-a-kind design, construction, and future initial operation of AP-1000 nuclear power plants in China.

The NRC also initiated the preparation of an information exchange arrangement with the regulatory authority of Thailand (Office of Atoms for Peace). NRC expanded engagement on establishing basic regulatory infrastructure needed for oversight of a nuclear power program with additional countries of Southeast Asia, including Malaysia and the Philippines. The agency also expanded engagement with international regulatory counterparts, including countries from South America, Africa and Asia on establishing effective regulatory oversight of uranium recovery activities and facilities.

In FY 2011, the NRC continued the program of assistance to Latin American countries in regulatory controls over radioactive materials, including the establishment or enhancement of national source registries, review of national legislation, and assistance in implementing a year-long regulatory staff qualification program. These activities were carried out in Chile, the Dominican Republic, Panama, Paraguay, and Uruguay.

Multilateral Cooperation and Assistance

The NRC held the second in a series of uranium recovery workshops for international counterparts through its international assistance program activities in May 2011. Regulatory bodies from the countries of Bolivia, Chile, Ecuador, Indonesia, Jordan, Mexico, Mongolia, Niger, Peru, Tanzania and Uruguay were represented. The focus of the workshop was to assist countries who are initiating or restarting uranium recovery regulatory programs. The overall goal of the workshop was to provide information on regulatory development, licensing, regulatory oversight, and prevention of legacy sites when uranium production ceases at a site. The next workshop is planned to be held in Tanzania and to invite representatives of the regulatory bodies from African countries who are initiating uranium recovery programs.

Multilateral Nuclear Safety Organizations

The NRC is engaged both domestically and internationally in efforts to enhance nuclear safety and security through the regulatory oversight of radioactive sources. In May 2011, the NRC participated in an IAEA meeting of technical and legal experts on the IAEA's Code of Conduct for the Safety and Security of Radioactive Sources to review and possibly revise the "Guidance on the Import and Export of Radioactive Sources" (the Guidance). The NRC also supported a July 2011 IAEA meeting for Member States, who have not yet made a political commitment to the Code and Guidance to explain the benefits of doing so. In addition, the agency continued radioactive source-related assistance to the countries of the Commonwealth of Independent States, expanded a provision of radioactive source-related assistance to include selected countries of Africa, Latin America, and Southeast Asia, conducted regional workshops on the physical protection of radioactive sources, and continued coordination with sources-related assistance provided by the IAEA and others. The agency also worked with other U.S. Government agencies, such as the Departments of State, Energy, and Commerce and the National Security Council Staff, and with the IAEA to develop international security guidance documents for radioactive sources.

The NRC continues to support the development and implementation of programs to leverage the knowledge and resources within the international regulatory community in the licensing of new reactor designs. The agency continued its leadership role in Multinational Design Evaluation Program (MDEP), through which regulatory authorities in 10 countries share expertise and resources in reviewing new reactor designs. The program consists of three issue-specific and two design-specific working groups. The Digital Instrumentation and Controls Working Group, led by the United States, drafted common positions in digital instrumentation and controls system design. The Vendor Inspection Cooperation Working Group conducted several parallel inspections that involved more than one regulator and the Codes and Standards Working Group is completing a project to compare the pressure boundary codes of five member countries. The design-specific working groups, based on the Westinghouse AP 1000 and the AREVA evolutionary power reactor designs, also established sub-working groups. In FY 2011, the Policy Group, which is the governing body of the program, modified the MDEP terms of reference to establish a process for additional countries to join. The revised terms of reference provides for two new types of membership, associate members and an MDEP candidate.

Chapter 2
PROGRAM PERFORMANCE

STRATEGIC GOAL 2: SECURITY

Ensure Adequate Protection in the Secure
Use and Management of Radioactive Materials

STRATEGIC OUTCOME

The NRC has the following strategic outcome associated
with its goal to ensure the secure use and management of
radioactive materials:

- Prevent any instances where licensed radioactive
  materials are used domestically in a manner hostile to
  the security of the United States.

The strategic outcome specifies the condition that must
be met for the agency to achieve its Security goal. In
FY 2011, the NRC achieved its Security goal strategic
outcome.

PERFORMANCE MEASURES

The NRC also uses annual performance measures
to assess whether the agency met its Security goal.
Performance measures are aligned at a lower risk level
than the strategic outcomes. As a result, not fully
achieving a performance measure may not cause harm
to the public or environment. Missing a performance
measure signals that safety levels may have deteriorated
at the agency strategic planning level. If the NRC misses
a performance measure, the agency will take corrective
actions to bring the measure back into the target range.
Table 4, on the next page, shows the agency's annual
performance measures.

Table 4 SECURITY GOAL PERFORMANCE MEASURES

Performance Measure	2006	2007	2008	2009	2010	2011
1. Number of unrecovered losses or thefts of risk-significant radioactive sources is zero.	0	0	0	0	0	1
2. Number of substantiated cases of theft or diversion of licensed, risk-significant radioactive sources or formula quantities of special nuclear material, or attacks that result in radiological sabotage, is zero.	0	0	0	0	0	0
3. Number of substantiated losses of formula quantities of special nuclear material or substantiated inventory discrepancies of formula quantities of special nuclear material that are caused by theft or diversion or by substantial breakdown of the accountability system sabotage is zero.	0	0	0	0	0	0
4. Number of substantial breakdowns of physical security or material control (i.e., access control containment or accountability systems) that significantly weaken the protection against theft, diversion, or sabotage is less than or equal to one.	0	0	0	0	0	0
5. Number of significant unauthorized disclosures of classified or safeguards information is zero.	0	0	0	0	0	0

Chapter 2
PROGRAM PERFORMANCE

1. *Unrecovered Losses or Thefts*

This measure tracks any loss or theft of radioactive nuclear sources that the NRC has determined to be of significant risk. The measure tracks the agency's performance in ensuring the proper accounting for radioactive sources of significant risk that could be used for malicious purposes. The ability to account for these sources is vital to securing the Nation's critical infrastructure from radiological crimes. There were no losses and one theft of radioactive nuclear material that the NRC determined to be risk-significant during FY 2011.

On July 19, 2011, in Austin, Texas, Licensee (Acuren Inspections, Inc.) notified Texas Department of Health that a truck had been broken into and that a radiography camera transportation container containing a QSA Global Model 880 D camera with a 33.7 currie iridium (Ir) 192 source and a portable electric generator had been stolen. For a detailed update, see Event Report No. 47070 on the NRC Website http://www.nrc.gov/reading-rm/doc-collections/event-status/event/2011/20110725en.html#en47070. The agency will coordinate and review the Increased Controls applied to these sources and determine if additional controls need to be implemented. If changes to the Increased Controls are needed, they will also be considered in an ongoing Part 37 Rulemaking.

2. *Thefts or Diversion*

This measure tracks whether NRC-licensed facilities maintain adequate protective capabilities to prevent theft or diversion of nuclear material or sabotage that could result in substantial harm to the public health and safety. There were no substantiated cases of theft or diversion of licensed, risk-significant radioactive sources or formula quantities of special nuclear material or attacks that resulted in radiological sabotage during FY 2011.

3. *Loss or Inventory Discrepancy*

This measure tracks whether special nuclear material is accounted for and that losses of this material do not occur that could lead to the creation of an improvised nuclear device or other type of nuclear device. The measure also tracks whether the systems in place at NRC-licensed facilities maintain accurate inventories of the special nuclear material that the facilities process, use, or store. There were no substantiated losses of formula quantities of special nuclear material or substantiated inventory discrepancies of formula quantities of special nuclear material that were caused by theft or diversion or by substantial breakdown of the accountability system during FY 2011.

4. *Substantial Breakdowns of Physical Security*

This measure tracks any breakdowns in access control, containment, or accountability systems that significantly weakened the protection against theft, diversion, or sabotage for nuclear materials the agency has determined to be of significant risk. There were no substantial breakdowns of physical security during FY 2011.

5. *Significant Unauthorized Disclosures*

This measure includes significant unauthorized disclosures of classified or Safeguards Information that cause damage to national security or public safety. This measure tracks whether information that can harm national security (Classified Information) or cause damage to the public health and safety (Safeguards Information) has been stored and used in such a way as to prevent its disclosure to the public, terrorist organizations, other nations, or personnel without a need to know. There were no significant disclosures that caused damage to national security or public safety during FY 2011.

Operations Center Exercise

NUCLEAR SECURITY PROGRAMS

The NRC must remain vigilant to protect the security of nuclear facilities and materials. The agency achieves its Security goal with licensing and oversight programs similar to those employed in achieving its Safety goal. The aim is to allow licensees to realize the benefits of nuclear materials through their secure use while placing only necessary regulatory requirements on them. The following sections describe the NRC's FY 2011 security activities that enabled the agency to achieve its Security goal, security strategic outcome, and security performance measures.

NEW AND OPERATING REACTOR SECURITY

The NRC conducts a robust security inspection program within the Security Cornerstone of the agency's Reactor Oversight Process. The Security Cornerstone focuses on five key attributes of licensee performance: access authorization, access control, physical protection systems, material control and accounting, and response to contingency events. Through the results obtained from all oversight activities, including baseline security inspections and performance indicators, the agency determines whether licensees are in compliance with NRC

requirements and can provide high assurance of adequate protection against the design-basis threat for radiological sabotage. There were no substantial breakdowns of physical security at any commercial nuclear power plant in FY 2011.

The NRC regularly carries out force-on-force inspections at commercial operating nuclear power plants and Category I fuel facilities as part of its comprehensive security program. The agency uses these inspections to evaluate the effectiveness of security programs to prevent radiological sabotage and theft or diversion of Category I material. The agency conducts force-on-force inspections at least once every three years at each commercial nuclear power plant and Category I fuel facility. Force-on-force inspections assess the ability of nuclear facilities to defend against the applicable design-basis threat, which characterizes the adversary against which licensees must design appropriate defenses, such as physical protection systems and response strategies. A force-on-force inspection includes tabletop drills and simulated combat between a mock commando-type adversary force and the site security force. During the attack, the adversary force attempts to reach and damage key safety systems and components at a nuclear power plant, steal material at a Category I fuel facility, or gain control of safeguarded material. In FY 2011, the agency completed 23 force-on-force inspections and three force-on-force reinspections at nuclear power plants.

The NRC continued to assess and address enhancements of current practices for granting unescorted access at nuclear power plants during FY 2011. The agency's activities included: amending the behavioral observation program, enhancing office procedures for coordination with the Terrorist Screening Center, and improving access to the Personnel Access Data System. The agency anticipates that licensees will voluntarily implement the above modifications to their Behavioral Observation Program in FY 2011. The NRC is in the process of installing terminals at its headquarters for direct access to the industry's information sharing database.

Chapter 2
PROGRAM PERFORMANCE

The NRC continued the enhancement of Fitness-For-Duty policy and technical support of FFD-related rulemaking, licensing, and oversight of drug and alcohol requirements for all commercial power reactor and Category 1 fuel cycle licensees, and other groups (such as new reactor construction entities and contractor/vendors). A rulemaking was completed to provide licensees subject to 10 CFR Part 26, Subpart I, an alternative method to manage work hours for persons performing safety- or security-related activities, in lieu of the original method that required calculating work hours based on the number of hours worked and number of days off, the alternate method establishes a simpler and more flexible 54-hour maximum average work-hour limit, calculated over a 6-week period, to help preclude cumulative fatigue.

SPENT FUEL, FUEL CYCLE FACILITY, AND TRANSPORTATION SECURITY

The NRC completed its FY 2011 core security inspection program at NRC-licensed materials and waste facilities and fuel cycle facilities. It also completed six site visits to review licensee implementation of the Independent Spent Fuel Storage Installations security orders.

In FY 2011, the NRC continued its efforts to establish and monitor classified information security programs for uranium enrichment vendors and mixed-oxide facilities, including readiness reviews at multiple fuel cycle facilities. These reviews included evaluation of physical and information system security at these sites, licensee contractors performing classified work, and foreign ownership, control, or influence considerations in support of the facility clearance. In addition, NRC personnel participated in Quadripartite Working Group and DOE meetings on the protection of sensitive information associated with the URENCO USA enrichment facility.

The NRC continued security rulemaking activities to enhance its security requirements for licensees. The agency published a proposed rule that would add a new Part 37, "Physical Protection of Byproduct Material," to Title 10 CFR, "Energy," and made conforming changes to other parts of 10 CFR. The rule will put in place generally applicable requirements for licensees that possess International Atomic Energy Agency (IAEA) Category 1 and Category 2 radioactive materials. The proposed rule addresses physical protection at the facilities during transit, as well as access to materials. The agency developed a draft technical basis for 10 CFR Part 73 rulemaking that focuses on fuel cycle facility security and considers material attractiveness and domestic and international stakeholder views. A proposed draft rulemaking is scheduled to be published for comment in May 2012. The agency is also engaged in reviewing stakeholder comments from a draft technical basis for Independent Spent Fuel Storage Installation security rulemaking. The NRC anticipates that these technical bases will support the commencement of rulemaking activities in these areas during FY 2012.

NUCLEAR MATERIAL USERS SECURITY

The NRC continued its efforts to mitigate the potential risk of terrorist threats through enhanced security and controls for the use, storage, and transportation of risk significant byproduct material and spent nuclear fuel. In collaboration with the Department of Homeland Security (DHS), DOE, and other Federal, State, and local agencies, the NRC continued to assess the potential use of risk-significant sources in radiological dispersal devices and to coordinate efforts to enhance radioactive source protection and security. The NRC also worked with Agreement States to implement requirements for licensees that enhance the security and control of risk-significant radioactive material, including development of an inspection program to verify the implementation of these measures.

The NRC staff participated in activities related to the Government Coordinating Council, which enables interagency and cross-jurisdictional coordination on critical infrastructure and key resources, including transportation and material security. The staff also participated in trilateral meetings with DHS and DOE National Nuclear Security Administration to enable coordination among the participants on issues related to radioactive material security.

CONTROL OF RADIOACTIVE SOURCES

The NRC also implemented the National Source Tracking Rule, which requires licensees to report information on the possession of IAEA Category 1 and 2 radioactive sources (i.e., nationally tracked sources). The rule requires NRC and Agreement State licensees to report transactions involving the manufacture, transfer, receipt, disassembly, and disposal of nationally tracked sources. In FY 2011, licensees completed the second annual inventory reconciliation of their nationally tracked sources.

The National Source Tracking System, and the future Web-based Licensing System and License Verification System, are key components of a comprehensive program for the security and control of radioactive material. The NRC is integrating all three systems into a common system environment and architecture to form an integrated source management system that will include information on all U.S. licensees and over 70,000 risk-significant radioactive sources possessed by approximately 1,400 licensees. The integrated system will provide licensees, regulators, and Federal agencies with an additional round-the-clock means of determining the legitimacy of individuals possessing or seeking to obtain radioactive material to ensure that the materials are obtained only in authorized amounts by legitimate users.

INTERNATIONAL SECURITY

During FY 2011, the agency issued 175 licenses for the export or import of Category 1 and Category 2 radioactive materials as defined by the Code. The NRC continued its significant participation in implementing portions of the IAEA Code of Conduct on the Safety and Security of Radioactive Sources, as well as its participation in IAEA committees that are the nuclear security series fundamentals, recommendations and guidance documents.

One of the most notable accomplishments was publication of INFCIRC/225, Revision 5, "Nuclear Security Recommendations on Physical Protection of Nuclear Materials and Nuclear Facilities." After nearly five years of global discussions, the agency's involvement in these committees enhances security and public safety and contributes to international and domestic regulatory consistency. The agency also participated, as part of a U.S. interagency team, in a number of visits to other countries in an effort to ensure that U.S.-origin nuclear material is receiving adequate physical protection in accordance with bilateral agreements.

In FY 2011, the NRC is finalizing a rulemaking change to its regulations pertaining to the export and import of nuclear materials and equipment. The rule change is necessary to reflect the nuclear, non-proliferation policy of the Executive Branch including U.S. Government obligations to the IAEA and its publication of INFCIRC/225/Revision 5.

INTEGRATED AND COORDINATED SECURITY ACTIVITIES

The NRC has working relationships with the Federal Bureau of Investigation (FBI), DHS, Nuclear Energy Institute (NEI), power reactor licensees, and State and local law enforcement agencies to create integrated approaches to security within the nuclear sector. One significant outcome is the Integrated Pilot Comprehensive

Exercise (IPCE). The IPCE is a voluntary, collaborative effort led by the FBI with the support of DHS, the NRC, and NEI. The IPCE incorporates Federal, State, and local law enforcement tactical response planning and operations into the concept of integrated response by providing law enforcement tactical teams with opportunities to prepare for and respond to simulated security incidents inside commercial nuclear power plants. An IPCE was conducted in June, 2011.

The NRC participated in many other nuclear sector activities under DHS's National Infrastructure Protection Plan framework, such as the Government Coordinating Council, Critical Infrastructure Partnership Advisory Council, Federal Senior Leadership Council, and Research and Development Working Group. The NRC also contributed to national policy documents, including the Nuclear Sector-Specific Plan, Nuclear Sector Critical Infrastructure and Key Resources Protection Annual Report, and the National Critical Infrastructure and Key Resources Annual Report.

CYBER SECURITY

The NRC issued 10 CFR 73.54, "Protection of Digital Computer and Communication Systems and Networks," in March 2009. Licensees and COL applicants are required to provide high assurance that nuclear power plant safety, security, and emergency preparedness functions are adequately protected from cyber attacks up to and including the design-basis threat.

In October 2010, the Commission determined, as a matter of policy, that the NRC's cyber security rule, 10 CFR 73.54, should be interpreted to include structures, systems, and components in the Balance of Plant that have a nexus to radiological health and safety at NRC-licensed nuclear power plants. In late 2010 leading into 2011, the agency developed a Standard Review Plan (SRP) used to add consistency to the evaluation of licensee-submitted cyber-security plans and implementation schedules. Using the SRP, the staff has approved cyber-

security plans for all commercial nuclear reactor licensees and is now in the process of drafting a temporary instruction and developing a regulatory oversight program for cyber-security. The inspection program is scheduled to begin in FY 2012.

COSTING TO GOALS

The NRC is working to improve its cost management capabilities to better align its costs with desired outcomes. This year's Performance and Accountability Report presents the full cost of achieving the safety and security goals for the agency's programs, Nuclear Reactor Safety and Security and Nuclear Materials Safety and Security. The cost of achieving the agency's safety goal was $1,024.0 million, and the cost of achieving the agency's security goal was $72.9 million (see Figure 9).

Figure 9
NRC SAFETY AND SECURITY COSTS
(In Millions)

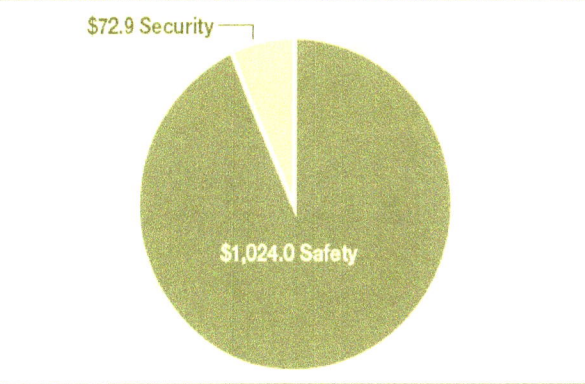

$72.9 Security

$1,024.0 Safety

ORGANIZATIONAL EXCELLENCE OBJECTIVES

The NRC has three Organizational Excellence Objectives: openness, effectiveness, and operational excellence. These objectives are critical components to carrying out the agency's regulatory mandate to serve the American people.

Chapter 2 PROGRAM PERFORMANCE

OPENNESS

The openness objective explicitly recognizes that the public must be informed about, and have a reasonable opportunity to participate in, the NRC's regulatory processes. The NRC is firmly committed to transparency, participation, and collaboration as key principles governing the agency's relationship with the public and other stakeholders. The agency has demonstrated its commitment to these openness principles through its long-standing efforts to keep stakeholders informed and involved in the NRC's regulatory process.

The NRC's response to the Open Government Directive reaffirms that commitment, extending agency efforts through the use of social media that enable rapid communication and interaction with stakeholders and collaboration technologies such as Web-conferencing tools that broaden participation in public meetings. In FY 2011, the agency has continued to implement its Open Government Plan (published in FY 2010 and available on the NRC Website http://www.nrc.gov/public-involve/open.html) and has been successful in meeting the plan's commitment.

Virtual meeting services were expanded to enable the agency to conduct a higher number of virtual meetings with NRC staff and external stakeholders. The NRC began its first major and sustained use of social media services by launching the NRC public blog. Additionally, the NRC launched a YouTube site and official Twitter channel. Further, 26 high interest public data sets were developed and published to Data.gov. These expanded capabilities helped the NRC to increase transparency, participation, and collaboration with the public. Additional activities supporting openness are described in the Information Technology and Information Management section.

Nuclear Reactor Safety

Operating Reactors

The topics of public meetings included fire protection (National Fire Protection Association Standard 805, "Performance-Based Standard for Fire Protection for Light-Water Reactor Electric Generating Plants"), license renewal reviews (at Crystal River, Seabrook, Davis-Besse, Salem/Hope Creek, and South Texas), B&W Medical Isotope Production Systems, and the medical isotope Molybdenum-99 shortage.

The NRC also held monthly public meetings during FY 2011 to discuss the Reactor Oversight Process. Participants discussed suggestions for improvement, questions, and program implementation issues. Additionally, the agency continued to provide accurate and timely information to the public by ensuring that non-sensitive, unclassified regulatory documents are released to the public by the sixth working day after the document date. The agency routinely holds public meetings to present the agency's assessments of safety performance at nuclear reactor sites.

The NRC maintains information on license renewal for commercial operating power reactors on its Website http://www.nrc.gov/reactors/operating/licensing/renewal.html. Processes, regulations, and inspection reports for the Reactor Oversight Process are also available on the the NRC Website http://www.nrc.gov/NRR/OVERSIGHT/ASSESS/index.html.

New Reactors

The NRC maintains project status and schedules for new reactor licensing activities monthly, making them available on the NRC Website http://www.nrc.gov/reactors/new-reactors.html. The NRC Website receives approximately 50,000 hits per month for information on new reactor licensing activities.

Chapter 2
PROGRAM PERFORMANCE

The NRC held over 140 public meetings on new reactor activities in FY 2011. These meetings engaged stakeholders in the regulatory process and provided information on public participation in the environmental review process. The agency actively solicited comments on the scope of environmental impact statements, and provided information on lessons learned about locating sites and environmental reviews.

The NRC staff also conducted numerous public meetings during FY 2011 to provide a forum for stakeholders to participate in and comment on staff proposals for ITAAC closure, ITAAC maintenance, and other construction inspection program issues. The agency also held public workshops on proposed rulemaking activities for design certification rule templates and ITAAC maintenance.

Nuclear Materials and Waste Safety

The NRC continued its active participation in many meetings to inform the public of its activities. Agency representatives attended meetings for the Institute of Nuclear Materials Management Spent Fuel Seminar, regional meetings of the Council of State Governments, the U.S. Transport Council, and the NEI Dry Cask Storage Forum on radioactive material transportation and spent fuel storage matters.

In its continuing efforts to reach out to stakeholders, the NRC conducted its sixth annual Fuel Cycle Information Exchange conference in June 2011. The Fuel Cycle Information Exchange addresses a broad range of issues in the licensing and oversight of new and operating fuel facilities and potential developments for future reactors and fuel cycles. It provides a forum for presentations and panel discussions involving regulators, industry, and public stakeholders, both domestic and international.

Materials

In FY 2011, the agency published the Policy Statement of the U.S. Nuclear Regulatory Commission on the Protection of Cesium-137 Chloride Sources. The policy statement was developed with extensive stakeholder input obtained from two public meetings and from comments received in response to publication of the draft statement in the *Federal Register*. The policy statement was published on July 25, 2011, and is accessible on the Website http://www.regulations.gov. The agency held two public meetings to engage the broader medical stakeholder community to develop event definitions that will protect patients and allow physicians the flexibility to take appropriate medical actions. These medical stakeholder meetings also considered other concerns with the medical regulations.

Decommissioning and Low-Level Waste

In FY 2011, the agency held 30 technical meetings with decommissioning licensees; uranium recovery facility applicants and licensees; and low-level waste stakeholders that were open to the public. The agency also engaged in outreach and consultation with Native American Tribes as part of efforts to fulfill the agency's Section 106 responsibilities under the National Historic Preservation Act (NHPA). The agency collaborated with State Historic Preservation Officers, conducted independent research, and contacted other consulting parties to determine which Tribes may have an interest in uranium recovery licensing activities. The agency worked with interested Tribes to identify potentially affected sites and to assess and resolve any adverse effects to those sites related to uranium recovery licensing activities.

EFFECTIVENESS

The drive to improve performance in government, coupled with increasing demands on the NRC's resources, requires the NRC to become more effective, efficient, and timely in its regulatory activities. The agency's effectiveness initiatives sharpen the agency's focus on safety and security and ensure that its available resources are optimally directed toward accomplishing the agency's mission. In FY 2011, the agency performed a comprehensive review of NRC overhead functions (e.g., administrative services, human capital, financial management including contract management, information management, and information technology) to identify effective, efficient, and cost conscious business solutions and eliminate duplicative processes and functions.

Nuclear Reactor Safety

Operating Reactors

In October 2010, the IAEA conducted an Integrated Regulatory Review Service (IRRS) mission-to-peer review of the NRC's operating power reactor program. The IRRS mission report included two recommendations and 20 suggestions which the NRC is considering, as well as a number of good practices. The NRC will also host an IAEA International Physical Protection Advisory Service (IPPAS) in 2013 to review protection at research and test reactors.

In June 2011, the NRC issued its final safety culture policy statement that sets forth expectations that individuals and organizations involved in NRC-regulated activities establish and maintain a positive safety culture proportionate to the safety and security significance of their activities. The statement reinforces the NRC's emphasis on a "safety-first" focus but is not a regulation and, as such, does not impose requirements. Safety culture refers to an organization's collective commitment, by leaders and individuals, to emphasize safety as an overriding priority to competing goals and other considerations to ensure protection of people and the environment. The policy statement complements agency regulations and guidance.

New Reactors

For the new reactor license applications currently under review, the NRC continued the use of earned value management project health indicators during FY 2011 to determine overall project health, improve schedule compliance and resource utilization, and improve the efficiency of the project under review. As a result of implementing earned value management, the agency increased the effectiveness of new reactor licensing in three ways. First, it focused limited resources on the new reactor projects that are expected to complete licensing and construction and begin operation in the near term. Second, it identified and minimized risks to project schedules and review completions. Third, it managed resource use across many complicated new reactor licensing applications.

Nuclear Materials and Waste Safety

The agency developed a plan for integrating spent nuclear fuel regulatory activities to more effectively address the regulatory and licensing aspects of extended storage and transportation (i.e., greater than 120 years), reprocessing, and disposal of spent nuclear fuel and high-level waste. The purpose of the plan is to ensure that the regulation of the back end of the fuel cycle accomplishes safety, security, and environmental protection in an efficient and effective manner and that decisions made about one component or area of this system adequately consider other components or areas (i.e. treating spent fuel and high-level waste regulation as a system of interrelated activities). By coordinating the approach for regulation of spent nuclear fuel or high-level waste storage, potential reprocessing, transportation, and disposal, the agency can improve the efficiency and effectiveness of NRC regulatory processes and provide stability and predictability for stakeholders in a dynamic environment.

The NRC proceeded with the revisions to the Consolidated Guidance Series (NUREG-1556) to address incorporation of security issues and update licensing practices which will enhance the materials licensing review process.

As part of the NRC's license review process, the agency performs an acceptance review to determine if the license application contains adequate information. To aid the environmental review of uranium recovery in-situ leach applications, the agency continues to tier-off from the Generic Environmental Impact Statement. Following the licensing of the first three new uranium recovery facilities, the agency held lessons-learned meetings with stakeholders and Tribal Representatives regarding the safety and environmental license review process. The purpose of those meetings was to identify improvements to make the review process more effective in the future for the agency, applicants, stakeholders, and Tribes.

OPERATIONAL EXCELLENCE

This objective focuses on the activities related to financial management, management of human capital, infrastructure management, and information technology and information management.

Financial Management

The NRC made substantial progress in modernizing its financial systems in FY 2011. On October 1, 2010, the NRC deployed the Financial Accounting and Integrated Management Information System (FAIMIS) Core Financial System (CFS). The FAIMIS CFS modernizes the agency's Core Financial Systems and consolidates five legacy systems, including nine FEES subsystems, within a single business solution. The FAIMIS CFS consolidates real-time financial information within a single modern Web-enabled system solution and provides modern and flexible reporting tools necessary to extract financial information from FAIMIS. Consolidated, accurate, and

consistent data, combined with the means to report on the data, has improved financial decision making and financial management significantly.

NRC FEES Panel at the Regulatory Information Conference

In FY 2011, the NRC continued its excellence in financial reporting. For the eighth consecutive year, an independent auditor has rendered an unqualified opinion on the NRC financial statements. The auditor also rendered an unqualified opinion on the agency's internal controls concluding that the NRC had no reportable conditions or significant deficiencies. In addition, the NRC received its tenth consecutive *Certificate of Excellence in Accountability Reporting* from the Association of Government Accountants (AGA), as well as AGA's *Best-In-Class Award* for providing the most comprehensive and candid presentation of forward-looking information in its FY 2010 Performance and Accountability Report.

The NRC continued to make progress in implementing a 21st Century Strategic Acquisition Program, an integrated financial and acquisition planning, execution and reporting methodology based upon business process improvements through the implementation of leading practices and system modernization. The approach is based upon enterprise spending management and strategic sourcing principles that have a proven track record of success in industry and Federal agencies. The agency's first spending analysis was completed and a pilot Portfolio

Council is underway. The agency is also in the process of selecting an Acquisition System Module that will be integrated with the agency's core financial system. The system will provide a single acquisition portal and document generator with defined workflow, business rules and enhanced reporting capability. A data warehouse of the NRC's DOE Laboratory and Interagency (IA) Agreements, along with their critical supporting documentation has been established in anticipation of being migrated into the Acquisition System.

Management of Human Capital

For several years the NRC experienced significant growth resulting from an increased interest in nuclear power. Currently, agency staff and resource levels have stabilized and it is unlikely that there will be any growth over the next several years. This requires that the NRC adjust its human capital strategies to ensure its continued success. The NRC took steps to meet this challenge by institutionalizing an approach that focused on its mission of protecting public health and safety while remaining mindful of staff needs.

NRC Recruiting at College Campuses

Through the Human Capital Council and its associated subgroups, the NRC has fostered a strong cohesive partnership with agency offices to develop and implement human capital initiatives.

For example, the NRC restricted external hiring to only the most critical skill sets, while still emphasizing Governmentwide programs such as hiring of the disabled and employment of veterans. The NRC maximized internal movement to meet changing resource needs. With these objectives in mind, the agency hosted an internal career fair earlier this year, which gave staff an opportunity to explore career options involving rotations and reassignments. To continue to enhance diversity management and support the agency's Comprehensive Diversity Management Plan, the NRC developed action plans specific to each NRC office with the expressed purpose of implementing strategies to support the agency's goal to increase, develop, and retain diversity at all levels, especially at the managerial level. Additionally, the NRC developed and submitted a Disability Program Strategic Project Plan to OPM which outlined plans for recruiting and retaining individuals with disabilities, a focus area for recruitment. The NRC made significant progress towards achieving actions included in the NRC Veterans' Action Plan.

The NRC used a variety of human capital strategies to maintain and bolster knowledge and skills during a period when a large number of experienced staff members are becoming eligible to retire. For example, the NRC continued to enhance its Knowledge Management program by actively capturing lessons learned from subject matter experts, improving access to lessons learned and training programs and aggressively building an agencywide Knowledge Center.

The NRC continued to implement training technologies such as on-line and distance learning to deliver high quality learning products at a reduced cost. The NRC is on track to achieve a third annual increase in the percentage of training conducted on-line. In FY 2011, 75.1 percent of the 39,307 course completions were conducted as on-line courses. The NRC continued to improve the Learning Management System (i-Learn) resulting in a fourth consecutive semi-annual increase in overall user satisfaction. User satisfaction as of March 2011 was 89.3 percent, up from the baseline of 59 percent in December 2009. OPM recognized NRC's expertise in developing effective on-line courses by recommending that all Federal agencies use NRC's No Fear Act on-line course.

Based on the 2011 Federal Employee Viewpoint Survey (FEVS) results, the NRC maintained its status as a stand-out role model for Governmentwide Human Capital indices and key OPM initiatives including Leadership and Knowledge Management, Performance Culture, Talent Management, Job Satisfaction, Hiring Reform, Diversity and Telework. The NRC developed and implemented highly effective efforts to further improve the agency's response rate for the Federal Employee Viewpoint Survey. The NRC response rate for the 2011 FEVS was 69.1 percent, up from 67.3 percent in last year's FEVS. This contrasts favorably to the governmentwide average response rate of 49.3 percent for the 2011 FEVS. The employees once again rated the NRC number one governmentwide as measured by all four of the Human Capital Assessment and Accountability indices.

The NRC rolled out the Federal Hiring Reform Initiative ahead of schedule. As of October 2010, all NRC vacancy announcements that were open to the public no longer required narrative (long-answer) KSA questions. The other components of the President's initiative on Hiring Reform had already been implemented at the agency.

The NRC conducted outreach to potential grant recipient institutions to market the grants program. For FY 2011, the agency awarded $9.7 million through 44 grant awards to educational institutions. These grants assist in expanding the workforce in nuclear safety and nuclear-related disciplines and the development of the next generation nuclear workforce.

The NRC and the National Treasury Employees Union (NTEU) continue to have a working agency level partnership (ALMPC). NRC management and NTEU identified goals and metrics to measure partnership accomplishments and submitted this information to the National Council on Federal Labor Management Relations (National Council) as required. In December 2011, the ALMPC will report results against these metrics to the National Council.

The NRC participated in several benchmarking studies with other agencies. For example, given the NRC's number one ranking in Leadership (as rated by the Partnership for Public Service), OPM requested that the NRC share its practices on training and developing its leaders with other Federal agencies. In addition, the NRC participated in a benchmarking study with the Department of Commerce, components of DOE, and the Department of Veterans Affairs on human resource operations.

Infrastructure Management

Substantial progress was made on the construction of a new NRC headquarters building that will house approximately 1,350 NRC employees and contractors. The new building's 14 floor above ground concrete shell was completed, and installation of the building's exterior walls and windows has started. Construction is scheduled to be completed by the end of FY 2012.

The NRC has taken a number of actions to improve employee security at the White Flint Complex (WFC), including: expanding the lobby of the government-owned One White Flint North Building which will provide better visitor management and a more secure area to screen visitors and their packages; installing access control turnstiles with electronic badge authentication; and closing the pedestrian entrance to visitors entering the Two White Flint North Building. In addition, the NRC repaved the WFC plaza in an effort to improve pedestrian safety.

New NRC Headquarters Building Construction Site

In support of Executive Order 13514, Federal Leadership in Environmental, Energy, and Economic Performance, the NRC installed energy management system equipment and variable frequency drives on the heating and cooling system, new LED lighting on the White Flint Complex plaza, lighting controls in the garage, and an electrical conditioning system, all of which resulted in a reduction in electrical consumption. The agency also continued efforts in support of greening the government with the installation of motion sensor activated faucets, toilets, and urinals to reduce water consumption. In addition, for the second consecutive year, the NRC was recognized for Outstanding Achievement in Recycling by Montgomery County, MD.

Information Technology and Information Management

The NRC continued to identify opportunities to improve program performance and information availability through the use of information technology (IT) solutions. Progress continued in several major focus areas to achieve operational excellence through more effective information management, effective IT infrastructure, and continuous customer service improvements.

Effective information management ensures needed information is available to the staff to help support predictable regulatory programs and policies. It also allows the NRC to meet its openness objective related to informing and involving stakeholders in the regulatory process by providing timely access to accurate agency information. FY 2011 accomplishments in this area included: 1) modernizing the Agencywide Documents Access and Management System (ADAMS) to ensure staff and stakeholders can readily access needed information; 2) deploying improved public search initiatives, such as a single-search capability, that provide stakeholders with more thorough search results; 3) redesigning and modernizing the NRC's Website to help provide users with better organized and more easily navigable information; 4) continuing personal interactions with stakeholders through the Public Document Room where stakeholders can work directly with a person to retrieve information; 5) providing key information dissemination by issuing timely public meeting notices, FOIA responses, and documents made publicly available through ADAMS; and 6) deploying of Safeguards Information Local Area Network and Electronic Safe System.

Effective IT infrastructure ensures that the NRC has a reliable and responsive foundation of technology to support business needs and agency operations. A major achievement in this area was the award of the new Information Technology Infrastructure and Support Services contract, which will provide key IT services across the enterprise. Other key successes included: the transition of telecommunications services from FTS2001 to Networx; the implementation of the Verizon Notification System that provides important agency communications to staff member personal phones and email accounts on a voluntary basis; allowing network access through the use of Personal Identity Verification cards; the expansion of the Safeguards Information Local Area Network and Electronic Safe system to HQ satellite offices, regional offices, and all resident inspector sites to allow secure network access to safeguards information data; the replatforming and redesign of the Protected Web Server used to securely communicate security incidents that occur at licensee locations; and the replacement of the Headquarters voicemail system. Many additional accomplishments in this area focused around the theme 'working from anywhere' and included: implementation/expansion of loaner programs for laptops and mobile broadband cards; implementation of home use programs for Microsoft Office 2010 and Adobe Acrobat Professional; significantly increasing the number of Mobile Desktops (laptops for home and office use); increasing the functionality of Webmail with the implementation of MessageWare; implementation of Guest Network Services allowing visitors and vendors secure Internet access from within NRC facilities; and the agency-wide deployment of Adobe Acrobat Professional software.

Effective IT project management ensures that projects across the agency are brought to a successful conclusion by offering best practices for a variety of project management disciplines. In FY 2011, a Project Management Office function continued to provide expertise, support and outreach through education programs, technical project management best practices, and business analysis/process support services. Nine business analysis engagements supporting key business processes and IT initiatives for program offices were completed in FY 2011.

Another primary focus area is service, a key component of operational excellence across the agency. In a continuing effort to evaluate the effectiveness of its IT/information management (IM) services, the NRC solicited feedback from employees on its IT/IM program by adding questions on this topic to the employee viewpoint survey conducted in FY 2011. The agency is also conducting an independent validation and verification of its IT/IM services to assess and fully understand the costs of existing services and to establish clear service expectations. Communication plans have been developed to further improve agency understanding of IT/IM services, and performance expectations. The NRC has also conducted a facilitating process to update its IT/IM Strategic Plan in coordination with on-going efforts to update the NRC Strategic Plan. The IT Service Catalog listing IT services continues to be enhanced to provide additional information and to improve service request capabilities.

PROGRAM EVALUATIONS

The NRC conducted several program evaluations of its regulatory operations during FY 2011. The evaluations were conducted for both the nuclear reactor and the nuclear materials programs.

OPERATOR LICENSING PROGRAM

Before the NRC licenses an individual to operate or supervise the controls of a commercial nuclear power reactor, the applicant must complete extensive training and pass rigorous examinations. Once licensed, operators and senior operators must comply with a number of requirements to maintain and renew their licenses. In FY 2011, an agency review team evaluated the operator licensing programs of two regions for their overall effectiveness and adherence to the guidance contained in NUREG-1021, Revision 9, *Operator Licensing Examination Standards for Power Reactors*, issued in July 2004, and other policy documents. The operator licensing programs are broken down into seven functional areas that are rated as either "satisfactory," or "needs improvement." The review team found the operator licensing programs in the two regions to be in accordance with the examination standards and assessed all areas as satisfactory. The review team also commended the regions' efforts to improve the quality of their examination packages.

REACTOR OVERSIGHT PROGRAM

The NRC completed a self-assessment of the Reactor Oversight Process in April 2011. The report, "Reactor Oversight Process Self-Assessment for Calendar Year 2010" (SECY-10-0042), is available on the NRC Website. The results of the calendar year 2010 self-assessment indicated that the Reactor Oversight Process met its program goals and achieved its intended outcomes. The Reactor Oversight Process was found to be objective, risk-informed, understandable, and predictable, and it met the agency goals of ensuring safety, openness, and effectiveness. The agency maintained its focus on stakeholder involvement and continued to improve the Reactor Oversight Process. The agency implemented improvements to address issues that were raised internally, recommended by independent reviews, and obtained from internal and external stakeholder feedback.

The NRC inspection and assessment program independently verified that nuclear power plants were operated safely and securely. The assessment program was revised to incorporate lessons learned from implementation of the safety culture enhancements and continued to ensure that the staff and licensees acted as necessary to address identified performance issues. The agency continues to improve the performance indicator program to ensure that the performance indicators are meaningful inputs to the Reactor Oversight Process, and it actively solicits input from internal and external stakeholders to further improve the Reactor Oversight Process based on stakeholder feedback and lessons learned.

INTEGRATED MATERIALS PERFORMANCE EVALUATION PROGRAM REVIEWS OF SELECTED NRC REGIONAL OFFICES

The NRC evaluates its own regional materials programs and Agreement State radiation control programs using performance indicators to ensure that public health and safety is adequately protected. The NRC, with the assistance of the Agreement States, completed ten Integrated Materials Performance Evaluation Program reviews to determine the adequacy and compatibility of the programs in the evaluated Agreement States during FY 2011. No regional evaluations were conducted during FY 2011.

DATA SOURCES, DATA QUALITY, AND DATA SECURITY

The NRC's data collection and analysis methods are driven largely by the regulatory mandate that Congress entrusted to the agency. Specifically, the NRC's mission is to regulate the Nation's civilian use of byproduct, source, and special nuclear materials to ensure adequate protection of public health and safety, protect the environment, and promote the common defense and security. In undertaking this mission, the agency oversees nuclear power plants, nonpower reactors, nuclear fuel facilities, interim spent fuel storage, radioactive material transportation, disposal of nuclear waste, and the industrial and medical uses of nuclear materials.

As part of the agency's regulatory requirement under 10 CFR 20.2206, several NRC-regulated industries are required to submit occupational radiation exposure reports to the Radiation Exposure Information and Reporting System database. NRC staff conducts analysis of these reports to ensure that NRC licensees comply with the annual occupational dose limit of 50 mSv (5 rem). NRC staff use the data in the following ways: (1) as one metric in the agency's Reactor Oversight Program to evaluate the effectiveness of licensee programs used to maintain occupational radiation doses as low as reasonably achievable and for inspection planning; (2) to assist in the evaluation of the radiological risk associated with certain categories of NRC-licensed activities and for comparative analysis of radiation protection performance; (3) to provide occupational radiation exposure history reports to individuals exposed to radiation or radioactive material at NRC-licensed facilities; and (4) to provide facts for responding to Congressional and Administration inquiries and to questions from the public regarding occupational radiation exposures at NRC-licensed facilities. The NRC publishes NUREG-0173, "Occupational Radiation Exposure at Commercial Nuclear Power Reactors and Other Facilities," annually. NUREG-0173 Volume 31 for Calendar Year 2009, was issued May 2011. It is available on the agency's Website http://pbadupws.nrc.gov/docs/ML1108/ML110820543.pdf.

Section 208 of the *Energy Reorganization Act of 1974*, as amended, requires the NRC to inform Congress of incidents or events that the Commission determines to be significant from the standpoint of public health and safety. The agency developed the Abnormal Occurrence Criteria to comply with the legislative intent of the Energy Reorganization Act to determine which events should be considered significant. Based on these criteria, the agency prepares an annual, "Report to Congress on Abnormal Occurrences," (NUREG-0090). One important characteristic of this report is that the data presented normally originate from external sources, such as Agreement States and NRC licensees. NUREG-0090 Volume 33 for FY 2010, issued June 2011, is available on the agency's Website http://www.nrc.gov/reading-rm/doc-collections/nuregs/staff/sr0090/v33/sr0090v33.pdf.

The NRC finds these data sources credible because (1) agency regulations require Agreement States, licensees,

and other external sources to report the necessary information, (2) the NRC maintains an aggressive inspection program that, among other activities, includes auditing licensee programs and evaluating Agreement State programs to ensure that they are reporting the necessary information as required by the agency's regulations, and (3) the NRC has established procedures for inspecting and evaluating licensees. The agency employs multiple database systems to support this process, including the Licensee Event Report Search System, the Accident Sequence Precursor Database, the Nuclear Materials Events Database, and the Radiation Exposure Information Reporting System. In addition, non-sensitive reports submitted by Agreement States and NRC licensees are available to the public through ADAMS, accessible through the agency's Website http://www.nrc.gov/reading-rm/adams.html.

The NRC verifies the reliability and technical accuracy of event information reported to the agency. The agency periodically inspects licensees and reviews Agreement State programs. In addition, NRC Headquarters, the regional offices, and Agreement States hold periodic conference calls to discuss event information. Events identified as meeting the Abnormal Occurrence Criteria are validated and verified before being reported to Congress.

Additionally, the NRC is an active participant in Data.gov, a Federal Website designed to increase public access to high-value, machine-readable datasets generated by the Executive Branch. The NRC published its first dataset in October 2009, and in response to the Open Government Directive published three additional datasets in January 2010. The NRC will continue to encourage public feedback on its high-value information, and consistent with agency policy and guidance provided by Data.gov, will continue to add new datasets to its high-value dataset publication plan.

INFORMATION SECURITY

The NRC's information security program (1) protects NRC and licensee information and information systems from unauthorized access, use, disclosure, disruption, modification, or destruction, (2) protects electronic control functions from unauthorized access or manipulation, and (3) ensures that adequate controls for protecting security-related information are used in the conduct of NRC business. The NRC information security program includes measures to accomplish the following:

(1) Ensure that information security requirements, standards, and guidance are clear, concise, appropriate, and able to mitigate the potential adverse effects if sensitive information is compromised.

(2) Ensure that security controls for information owned by or under the control of the NRC are consistent with established information security controls, that security controls for information are operating as intended and that they are having the desired impact, and that similar controls for licensees regulated by the NRC are in compliance with NRC information security regulations.

(3) Ensure that suspected or actual information security violations are evaluated and appropriate sanctions are considered.

(4) Ensure that the NRC has made sufficient preparations for information security-related emergencies and incidents.

(5) Ensure that internal information security program components complement each other and are periodically evaluated and improved.

PERFORMANCE DATA COMPLETENESS AND RELIABILITY

In order to manage for results, it is essential that the NRC assess the completeness and reliability of its performance data. Comparisons of actual performance with the projected levels are possible only if the data used to measure performance are complete and reliable. Consequently, the *Reports Consolidation Act of 2000* requires the NRC Chairman to assess the completeness and reliability of the performance data used in this report. The process for ensuring that the data are complete and reliable requires offices to complete a template for submission to the Chief Financial Officer for every performance measure certifying the data submitted have been approved by the applicable office director. The report "Verification and Validation of NRC's Performance Measures and Metrics" contains the processes the agency uses to collect, validate, and verify performance data. This report can be found in Appendix III of the NRC's FY 2011 Congressional Budget Justification located on the NRC Website http://www.nrc.gov/reading-rm/doc-collections/nuregs/staff/sr1100/v26/sr1100v26.pdf.

DATA COMPLETENESS

The NRC considers data to be complete if the agency reports actual performance data for every performance goal and indicator in the annual plan. Actual performance data include all data that are available when the agency sends its report to the President and Congress. The agency has reported actual data for every strategic and performance goal measure. In addition, all of the data is reported for each measure. As a result, the data presented in this report meet the requirements for data completeness.

DATA RELIABILITY

The NRC considers data to be reliable when agency managers and decision-makers use the data in carrying out their responsibilities. The data presented in this report meet this requirement for data reliability because NRC managers and senior leaders regularly use the reported data in the course of their duties.

Chapter 3

FINANCIAL STATEMENTS AND AUDITOR'S REPORT

A MESSAGE FROM THE CHIEF FINANCIAL OFFICER

I am pleased to present the financial statements for the U.S. Nuclear Regulatory Commission (NRC) Fiscal Year (FY) 2011 Performance and Accountability Report. For the eighth consecutive year, an independent auditor has rendered an unqualified opinion on the NRC financial statements. The auditor also rendered an unqualified opinion on our internal controls concluding that the NRC had no reportable conditions or significant deficiencies.

FY 2011 was financially challenging for the NRC. On October 1, 2010, the NRC successfully transitioned from our legacy core financial system that consisted of five stand alone systems with nine subsystems to an externally hosted integrated core financial system. However, production challenges with the new system delayed issuing reports from the cost accounting and fee billing modules, necessitating enhanced administrative controls to ensure data quality for financial operations and reporting. Additionally, the NRC budget execution was challenged during the year by emerging work associated with the agency response to the nuclear accident in Japan during the period of extended Continuing Resolution Appropriations. Agency actions were successful in meeting mission needs and maintaining a clean opinion on our financial statements, but some of our planned financial system improvements were delayed.

In FY 2012, we expect the challenging financial situation to continue as we adjust agency workload to implement the Japan Nuclear Accident Lessons Learned improvements under the government-wide tight budgetary conditions. The NRC plans to continue its financial system modernization to enhance financial operations and streamline agency operations. We plan to re-host the core financial system and continue development of system enhancements to seamlessly align budget development and execution functions. We also plan to modernize our Time and Labor System at the beginning of the fiscal year to improve its usability. We will also update the NRC Strategic Plan to set clear high level direction and goals for the agency in accordance with the *Government Performance and Results Modernization Act of 2010*. The new Strategic Plan will improve the link between the NRC budget structure and strategies for accomplishing our mission.

The NRC is committed to ensuring the safety and security of the Nation's civilian use of nuclear materials in the most effective and efficient manner. The regulation of the Nation's nuclear industry during this period of expansion and change requires rigorous stewardship of limited taxpayer resources and demands superior financial performance. I am proud of the progress we have made during the past year to promote sound business practices in the conduct of our regulatory mission and am confident that we will continue to make future improvements.

Pryer

J.E. Dyer
Chief Financial Officer
November 9, 2011

PRINCIPAL STATEMENTS

BALANCE SHEET *(In Thousands)*

As of September 30,	2011	2010
Assets		
Intragovernmental		
Fund balance with Treasury (Note 2)	$ **394,580**	$ 420,080
Accounts receivable (Note 3)	**8,287**	7,674
Other-Advances and prepayments	**3,681**	3,073
Total intragovernmental	**406,548**	430,827
Accounts receivable, net (Note 3)	**92,009**	123,242
Property and equipment, net (Note 4)	**46,542**	36,231
Other	**41**	25
Total Assets	$ **545,140**	$ 590,325
Liabilities		
Intragovernmental		
Accounts payable	$ **13,554**	$ 13,876
Other (Note 5)	**4,010**	5,986
Total intragovernmental	**17,564**	19,862
Accounts payable	**29,648**	26,666
Federal employee benefits (Note 6)	**7,245**	7,575
Other (Note 5)	**75,158**	106,041
Total Liabilities	**129,615**	160,144
Net Position		
Unexpended appropriations	**310,332**	311,869
Cumulative results of operations (Note 8)	**105,193**	118,312
Total Net Position	**415,525**	430,181
Total Liabilities and Net Position	$ **545,140**	$ 590,325

The accompanying notes to the principal statements are an integral part of this statement.

STATEMENT OF NET COST *(In Thousands)*

For the years ended September 30,	2011	2010
Nuclear Reactor Safety and Security		
Gross costs	$ 857,569	$ 882,591
Less: Earned revenue	(786,741)	(836,303)
Total Net Cost of Nuclear Reactor Safety and Security (Note 9)	70,828	46,288
Nuclear Materials and Waste Safety and Security		
Gross costs	239,350	257,862
Less: Earned revenue	(101,919)	(87,178)
Total Net Cost of Nuclear Materials and Waste Safety and Security (Note 9)	137,431	170,684
Net Cost of Operations	$ 208,259	$ 216,972

The accompanying notes to the principal statements are an integral part of this statement.

STATEMENT OF CHANGES IN NET POSITION *(In Thousands)*

For the years ended September 30,	2011	2010
Cumulative Results of Operations		
Beginning Balance	$ 118,312	$ 128,359
Budgetary Financing Sources		
Appropriations used (Note 11)	134,626	137,113
Non-exchange revenue (Note 11)	-	-
Transfers-in/out without reimbursement	9,980	29,000
Other Financing Sources		
Imputed financing from costs absorbed by others (Note 11)	50,534	40,812
Total Financing Sources	195,140	206,925
Net Cost of Operations	(208,259)	(216,972)
Net Change	(13,119)	(10,047)
Cumulative Results of Operations	$ 105,193	$ 118,312
Unexpended Appropriations		
Beginning Balance	$ 311,869	$ 338,637
Budgetary Financing Sources		
Appropriations received	133,346	128,345
Other adjustments (Rescissions)	(257)	(18,000)
Appropriations used (Note 11)	(134,626)	(137,113)
Total Budgetary Financing Sources	(1,537)	(26,768)
Total Unexpended Appropriations	310,332	311,869
Net Position	$ 415,525	$ 430,181

The accompanying notes to the principal statements are an integral part of this statement.

Chapter 3
FINANCIAL STATEMENTS AND AUDITOR'S REPORT

STATEMENT OF BUDGETARY RESOURCES *(In Thousands)*

For the years ended September 30,	2011	2010
Budgetary Resources		
Unobligated balance, brought forward, October 1	$ 44,699	$ 81,126
Recoveries of prior year unpaid obligations		
Actual	18,841	22,446
Budget authority		
Appropriation	1,054,219	1,066,859
Spending authority from offsetting collections		
Reimbursements earned-collected	12,439	10,086
Reimbursements earned-change in receivables	946	(424)
Change in unfilled customer orders-advance received	(3,506)	1,198
Change in unfilled customer orders-without advance	4,614	493
Subtotal-spending authority from offsetting collections	14,493	11,353
Permanently not available	(257)	(18,000)
Total Budgetary Resources	**$ 1,131,995**	**$ 1,163,784**
Status of Budgetary Resources		
Obligations incurred (Note 12)		
Direct	$1,078,667	$ 1,108,948
Reimbursable	4,818	10,137
Subtotal	1,083,485	1,119,085
Unobligated balance		
Apportioned	28,853	29,744
Exempt from apportionment	9,892	7,079
Subtotal	38,745	36,823
Unobligated balance, not available	9,765	7,876
Total Status of Budgetary Resources	**$1,131,995**	**$ 1,163,784**
Change in Obligated Balance		
Obligated balance, net		
Unpaid obligations brought forward, October 1	$ 375,381	$ 367,498
Obligations incurred, net	1,083,485	1,119,085
Gross outlays	(1,088,396)	(1,088,687)
Recoveries of prior year unpaid obligations, actual	(18,841)	(22,446)
Change in uncollected customer payments, from Federal sources	(5,560)	(69)
Obligated balance, net, end of period		
Unpaid obligations	359,402	383,154
Uncollected customer payments, from Federal sources	(13,333)	(7,773)
Total unpaid obligated balance, net, end of period	**$ 346,069**	**$ 375,381**
Net outlays		
Gross outlays	$1,088,396	$ 1,088,687
Offsetting collections	(8,933)	(11,284)
Distributed offsetting receipts	(910,901)	(909,514)
Net Outlays	**$ 168,562**	**$ 167,889**

The accompanying notes to the principal statements are an integral part of this statement.

NOTES TO THE PRINCIPAL STATEMENTS
(All Tables are Presented in Thousands)

Note 1.
SUMMARY OF SIGNIFICANT ACCOUNTING POLICIES

A. *Reporting Entity*

The U.S. Nuclear Regulatory Commission (NRC) is an independent regulatory agency of the Federal Government that was created by the U.S. Congress to regulate the Nation's civilian use of byproduct, source, and special nuclear materials to ensure adequate protection of the public health and safety, to promote the common defense and security, and to protect the environment. Its purposes are defined by the *Energy Reorganization Act of 1974*, as amended, along with the *Atomic Energy Act of 1954*, as amended, which provide the foundation for regulating the Nation's civilian use of nuclear materials.

The NRC operates through the execution of its congressionally approved appropriations for Salaries and Expenses (which includes funds derived from the Nuclear Waste Fund) and the Office of the Inspector General. In addition, the U.S. Agency for International Development (USAID) provides transfer appropriations to develop nuclear safety, regulatory authorities, and independent oversight of nuclear reactors in Russia, Ukraine, Kazakhstan, Georgia, and Armenia.

B. *Basis of Presentation*

These principal statements report the financial position and results of operations of the NRC as required by the *Chief Financial Officers Act of 1990* and the *Government Management Reform Act of 1994*. These financial statements were prepared from the books and records of the NRC in conformance with generally accepted accounting principles (GAAP) of the United States and the form and content for entity financial statements specified by the Office of Management and Budget (OMB) in Circular No. A-136, "Financial Reporting Requirements." GAAP for Federal entities are the standards prescribed by the Federal Accounting Standards Advisory Board, which is the official body for setting the accounting standards of the U.S. Government. These statements are, therefore, different from the financial reports, also prepared by the NRC pursuant to OMB directives, which are used to monitor and control the NRC's use of budgetary resources.

The NRC has not presented a Statement of Custodial Activity because the amounts involved are immaterial and incidental to its operations and mission.

C. *Budgets and Budgetary Accounting*

Budgetary accounting measures appropriation and consumption of budget spending authority or other budgetary resources and facilitates compliance with legal constraints and controls over the use of Federal funds. Under budgetary reporting principles, budgetary resources are consumed at the time of purchase. Assets and liabilities, which do not consume current budgetary resources, are not reported, and only those liabilities for which valid obligations have been established are considered to consume budgetary resources.

For the past 37 years, Congress has enacted no-year appropriations, which are available for obligation by the NRC until expended. The *Department of Defense and Full-Year Continuing Appropriations Act, 2011* requires the NRC to recover approximately 90 percent of its new budget authority by assessing fees for licensing and inspection activities.

D. *Basis of Accounting*

These financial statements reflect both accrual and budgetary accounting transactions. Under the accrual method, revenues are recognized when earned and expenses are recognized when a liability is incurred, without regard to receipt or payment of cash. Budgetary accounting is also used to record the obligation of funds prior to the accrual-based transaction. The Statement of Budgetary Resources presents budgetary resources available to the NRC and changes in obligations during the year. Interest on borrowings of the U.S. Department of the Treasury (Treasury) is not included as a cost to NRC programs and is not included in the accompanying financial statements.

E. Revenues and Other Financing Sources

The NRC is required to offset its appropriations by revenue received during the fiscal year from the assessment of fees. The NRC assesses two types of fees to recover its budget authority: (1) fees assessed under Title 10 of the *Code of Federal Regulations* (10 CFR) Part 170, "Fees for Facilities, Materials, Import and Export Licenses, and Other Regulatory Services under the *Atomic Energy Act of 1954*, as Amended," for licensing, inspection, and other services under the authority of the *Independent Offices Appropriation Act of 1952* to recover the NRC's costs of providing individually identifiable services to specific applicants and licensees; and (2) annual fees assessed for nuclear facilities and materials licensees under 10 CFR Part 171, "Annual Fees for Reactor Licenses and Fuel Cycle Licenses and Material Licenses." Licensing revenues are recognized on a straight-line basis over the licensing period. The annual licensing period for reactor and materials fees begins October 1 and ends September 30. Annual fees for reactors are invoiced in four quarterly installments, before the end of each quarter. The materials annual fee is invoiced in the month the license was originally issued. Inspection fees are recorded as revenues when the services are performed.

For accounting purposes, appropriations are recognized as financing sources (appropriations used) at the time goods and services are received. At the end of the fiscal year, appropriations recognized are reduced by the amount of assessed fees collected during the fiscal year to the extent of new budget authority for the year. Collections which exceed the new budget authority are held to offset subsequent years' appropriations. Appropriations expended for property and equipment are recognized as expenses when the asset is consumed in operations as reflected by depreciation and amortization expense.

F. Fund Balance with Treasury

The NRC's cash receipts and disbursements are processed by the Treasury. The Fund Balance with Treasury is primarily appropriated funds that are available to pay current liabilities and to finance authorized purchase commitments. Fund Balance with Treasury represents the NRC's right to draw on the Treasury for allowable expenditures.

G. Accounts Receivable

Accounts receivable consist of amounts owed to the NRC by other Federal agencies and the public. Amounts due from the public are presented net of an allowance for uncollectible accounts. The allowance is determined based on the age of the receivable and allowance rates established from historical experience. Receivables from Federal agencies are expected to be collected; therefore, there is no allowance for uncollectible accounts for Federal agencies.

H. Non-Entity Assets

Non-entity assets consist of miscellaneous penalties and interest due from the public, which, when collected, must be transferred to the Treasury.

I. Property and Equipment

Property and equipment consist primarily of typical office furnishings, leasehold improvements, nuclear reactor simulators, and computer hardware and software. The costs of internal use software include the full cost of salaries and benefits for agency personnel involved in software development. The NRC has no real property.

The land and buildings in which the NRC operates are provided by the General Services Administration (GSA), which charges the NRC rent that approximates the commercial rental rates for similar properties.

Property with a cost of $50 thousand or more per unit and a useful life of two years or more is capitalized at cost and depreciated using the straight-line method over the useful life. Other property items are expensed when purchased. Normal repairs and maintenance are charged to expense as incurred.

J. Accounts Payable

The NRC uses an estimation methodology to calculate the accounts payable balance which represents costs for billed and unbilled goods and services received (prior to year end) that are unpaid. The NRC had previously used an estimation methodology to calculate the accounts payable balance based on a review of the sample obligations from the total open obligations balances. For Fiscal Year 2011, the NRC calculates the accounts payable amount using an average based on historical trend of validated accruals. The estimation methodology is validated quarterly.

K. Liabilities Not Covered by Budgetary Resources

Liabilities represent the amount of monies or other resources that are likely to be paid by the NRC as the result of a transaction or event that has already occurred. No liability can be paid by the NRC absent an appropriation. Liabilities for which an appropriation has not been enacted are classified as "Liabilities Not Covered by Budgetary Resources." Also, the NRC liabilities arising from sources other than contracts can be abrogated by the Government acting in its sovereign capacity.

Intragovernmental

The NRC records a liability to the U.S. Department of Labor (DOL) for *Federal Employees Compensation Act* (FECA) benefits paid by DOL on behalf of the NRC.

Federal Employee Benefits

Federal employee benefits represent the actuarial liability for estimated future FECA disability benefits. The future workers' compensation estimate was generated by DOL from an application of actuarial procedures developed to estimate the liability for FECA, which includes the expected liability for death, disability, medical, and miscellaneous costs for approved compensation cases. The liability is calculated using historical benefit payment patterns related to a specific incurred period to predict the ultimate payments related to that period. These projected annual benefit payments are discounted to present value. The interest rate assumptions utilized for discounting benefits are 3.54 percent and 4.03 percent for FY 2011 and 3.65 percent and 4.22 percent for FY 2010.

Other

Accrued annual leave represents the amount of annual leave earned by NRC employees but not yet taken.

L. Contingencies

Contingent liabilities are those for which the existence or amount of the liability cannot be determined with certainty pending the outcome of future events. The NRC is a party to various administrative proceedings, legal actions, environmental suits, and claims brought by or against it. Based on the advice of legal counsel concerning contingencies, it is the opinion of management that the ultimate resolution of these proceedings, actions, suits, and claims will not materially affect the agency's financial statements. As of September 30, 2011, NRC was a party to one case where an adverse outcome was reasonably possible. The upper range of the loss on this potential liability is $150 thousand. As of September 30, 2010, the NRC was a party to one case where an adverse outcome was probable ($11.8 million) and one case where an adverse outcome was reasonably possible (upper range of $150 thousand). Treasury's Judgment Fund paid out on the $11.8 million FY 2010 contingent liability in FY 2011.

Chapter 3
FINANCIAL STATEMENTS AND
AUDITOR'S REPORT

M. Annual, Sick, and Other Leave

Annual leave is accrued as it is earned and the accrual is reduced as leave is taken. Each year, the balance in the accrued annual leave liability account is adjusted to reflect current pay rates. To the extent that current or prior year funding is not available to cover annual leave earned but not taken, funding will be obtained from future financing sources. Sick leave and other types of nonvested leave are expensed as taken.

N. Retirement Plans

The NRC employees belong to either the Federal Employees Retirement System (FERS) or the Civil Service Retirement System (CSRS). For FY 2011 and FY 2010, for employees belonging to FERS, the NRC withheld 0.8 percent of base pay earnings, in addition to *Federal Insurance Contributions Act* (FICA) withholdings, and matched the withholdings with an 11.5 percent contribution. The sum is transferred to the Federal Employees Retirement Fund. For employees covered by CSRS, the NRC withholds 7 percent of base pay earnings. The NRC matched this withholding with a 7 percent contribution in FY 2011 and FY 2010.

The Thrift Savings Plan (TSP) is a retirement savings and investment plan for employees belonging to either FERS or CSRS. The maximum percentage of base pay that an employee participating in FERS or CSRS may contribute is unlimited, subject to the maximum contribution of $16.5 thousand in 2011 and $16.5 thousand in 2010. For employees participating in FERS, the NRC automatically contributes one percent of base pay to their account and matches contributions up to an additional 4 percent. For employees participating in CSRS, there is no NRC matching of the contribution. The sum of the employees' and NRC's contributions are transferred to the Federal Retirement Thrift Investment Board.

The NRC does not report on its financial statements FERS and CSRS assets, accumulated plan benefits, or unfunded liabilities, if any, applicable to its employees. Reporting such amounts is the responsibility of the U.S. Office of Personnel Management. The portion of the current and estimated future outlays for CSRS not paid by the NRC is included in NRC's financial statements as an imputed financing source in NRC's Statement of Changes in Net Position and as program costs on the Statement of Net Cost.

O. Leases

The NRC's capital leases are for personal property consisting of reproduction equipment which is installed at NRC headquarters. For FY 2011, there are six capital leases with terms of five years, consisting of two capital leases added in FY 2011 with an interest rate of 1.26 percent, two capital leases added in FY 2008 with an interest rate of 3.99 percent, and two capital leases that were added in FY 2007 with an interest rate of 4.58 percent. The reproduction equipment is depreciated over five years using the straight-line method with no salvage value.

Operating leases consist of real property leases with GSA. The leases are for NRC's headquarters and regional offices. The GSA charges the NRC lease rates which approximate commercial rates for comparable space.

P. Pricing Policy

The NRC provides nuclear reactor and materials licensing and inspection services to the public and other Government entities. In accordance with OMB Circular No. A-25, "User Charges," and the *Independent Offices Appropriation Act of 1952*, the NRC assesses fees under 10 CFR Part 170 for licensing and inspection activities to recover the full cost of providing individually identifiable services.

The NRC's policy is to recover the full cost of goods and services provided to other Government entities where the services performed are not part of its statutory mission and the NRC has not received appropriations for those services. Fees for reimbursable work are assessed at the 10 CFR Part 170 rate with minor exceptions for programs that are nominal activities of the NRC.

Q. Net Position

The NRC's net position consists of unexpended appropriations and cumulative results of operations. Unexpended appropriations represent appropriated spending authority that is unobligated and has not been withdrawn by the Treasury and obligations that have not been paid. Cumulative results of operations represent the excess of financing sources over expenses since inception.

R. Use of Management Estimates

The preparation of the accompanying financial statements in accordance with Generally Accepted Accounting Principles requires management to make certain estimates and assumptions that affect the reported amounts of assets, liabilities, revenues, and expenses. Actual results could differ from those estimates.

S. Appropriation Transfers

The NRC is a party to allocation transfers with the USAID as a receiving (child) entity. These transfers are for the international development of nuclear safety and regulatory authorities in Russia, Ukraine, Kazakhstan, Georgia, and Armenia for the startup, operation, shutdown, and decommissioning of Soviet-designed nuclear power plants; the safe and secure use of radioactive materials; and the accounting for and protection of nuclear materials. Allocation transfers are legal delegations by one agency of its authority to obligate budget authority and outlay funds to another agency. All financial activity

related to these allocation transfers (e.g., budget authority, obligations, outlays) is reported in the financial statements of the parent entity from which the underlying legislative authority, appropriations, and budget apportionments are derived. The NRC receives allocation transfers, as the child, from USAID.

T. Statement of Net Cost

The programs as presented on the Statement of Net Cost are based on the annual performance budget and are described as follows:

The Nuclear Reactor Safety and Security Program encompasses all of the NRC efforts to ensure that civilian nuclear power reactor facilities and research and test reactors are licensed and operated in a manner that adequately protects the public health and safety, and the environment, and protects against radiological sabotage and theft or diversion of special nuclear materials. The Nuclear Reactor Safety and Security program contains the following activities: operating reactors and new reactors.

The Nuclear Materials and Waste Safety and Security program encompasses all NRC efforts to protect the public health and safety and the environment and ensures the secure use and management of radioactive materials. The Nuclear Materials and Waste Safety and Security program contains the following activities: fuel facilities, nuclear materials users, decommissioning and low-level waste, spent fuel storage and transportation, and high-level waste repository.

For intragovernmental gross costs, the buyers and sellers are both Federal entities. For earned revenues from the public, the buyers of the goods or services are non-Federal entities.

Note 2. FUND BALANCE WITH TREASURY

	2011	2010
Fund Balances		
Appropriated funds	$ 379,586	$ 400,435
Nuclear Waste Fund	15,098	19,645
Other fund types	(104)	-
Total	$ 394,580	$ 420,080
Status of Fund Balance with Treasury		
Unobligated balance		
Available		
Appropriated funds	$ 38,745	$ 36,823
Unavailable	9,765	7,876
Obligated balance not yet disbursed	346,069	375,381
Non-budgetary funds with Treasury	1	-
Total	$ 394,580	$ 420,080

The Fund Balance with Treasury consists of unobligated and obligated balance budgetary accounts. It includes Nuclear Waste Fund activity. The Nuclear Waste Fund unobligated balance is $9.9 million and $7.1 million as of September 30, 2011, and 2010, respectively.

Note 3. ACCOUNTS RECEIVABLE

	2011	2010
Intragovernmental		
Fee receivables and reimbursements	$ 8,287	$ 7,674
Receivables with the Public		
Materials and facilities fees-billed	$ 13,107	$ 2,611
Materials and facilities fees-unbilled	83,189	123,416
Other	180	77
Total Receivables with the Public	96,476	126,104
Less: Allowance for uncollectible accounts	(4,467)	(2,862)
Total Receivables with the Public, Net	$ 92,009	$ 123,242
Total Accounts Receivable	$ 104,763	$ 133,778
Less: Allowance for uncollectible accounts	(4,467)	(2,862)
Total Accounts Receivable, Net	$ 100,296	$ 130,916

Note 4. PROPERTY AND EQUIPMENT, NET

Fixed Assets Class	Service Years	Acquisition Value	Accumulated Depreciation and Amortization	2011 Net Book Value	2010 Net Book Value
Equipment	5-8	$ 12,942	$ (11,329)	$ 1,613	$ 1,941
Leased equipment	5-8	1,806	(1,157)	649	558
IT software	5	52,855	(42,798)	10,057	8,067
IT software under development	5	4,104	-	4,104	5,153
Leasehold improvements	20	44,437	(29,309)	15,128	14,040
Leasehold improvements in progress	-	14,991	-	14,991	6,472
Total		$ 131,135	$ (84,593)	$ 46,542	$ 36,231

Note 5. OTHER LIABILITIES

	2011	2010
Intragovernmental		
Liability to offset miscellaneous accounts receivable	$ 60	$ 6
Liability for advances from other agencies	81	82
Accrued workers' compensation	1,753	1,719
Accrued unemployment compensation	37	31
Employee benefit contributions	2,079	4,148
Total Intragovernmental Other Liabilities	$ 4,010	$ 5,986
Other Liabilities		
Accrued annual leave	$ 49,918	$ 50,413
Accrued salaries and benefits	9,138	26,621
Contract holdbacks, advances, capital lease liability, and other	5,344	7,391
Contingent liabilities	-	11,750
Grants payable	10,758	9,866
Total Other Liabilities	$ 75,158	$ 106,041
Total Intragovernmental and Other Liabilities	$ 79,168	$ 112,027

Other liabilities are current except for capital lease liability (Note 7).

Note 6. LIABILITIES NOT COVERED BY BUDGETARY RESOURCES

	2011	2010
Intragovernmental		
FECA paid by DOL	$ 1,753	$ 1,719
Accrued unemployment compensation	37	31
Federal Employee Benefits		
Future FECA	7,245	7,575
Other		
Accrued annual leave	49,918	50,413
Contingent liabilities	-	11,750
Total Liabilities not Covered by Budgetary Resources	58,953	71,488
Total Liabilities Covered by Budgetary Resources	70,662	88,656
Total Liabilities	$ 129,615	$ 160,144

Liabilities not Covered by Budgetary Resources represents the amount of future funding needed to pay the accrued unfunded expenses as of September 30, 2011, and 2010. These liabilities are not funded from current or prior-year appropriations and assessments, but rather should be funded from future appropriations and assessments. Accordingly, future funding requirements have been recognized for the expenses that will be paid from future appropriations.

Note 7. LEASES

	2011	2010
Assets under capital leases:		
Copiers and booklet maker	$ 1,806	$ 1,712
Accumulated depreciation	(1,157)	(1,154)
Net assets under capital leases	$ 649	$ 558

Future Lease Payments Due: Fiscal Year	Capital	Operating	2011	2010
2011	$ -	$ -	$ -	$ 31,647
2012	393	31,717	32,110	29,852
2013	105	33,696	33,801	24,754
2014	92	23,713	23,805	10,546
2015	93	20,781	20,874	7,603
2016 and thereafter	-	79,483	79,483	31,595
Total Lease Liability	683	189,390	190,073	135,997
Add: Imputed Interest	17	-	17	27
Total Future Lease Payments	$ 700	$ 189,390	$ 190,090	$ 136,024

The Capital Lease Liability of $683 thousand is included in Other Liabilities (Note 5).

Note 8. CUMULATIVE RESULTS OF OPERATIONS

	2011	2010
Liabilities not covered by budgetary resources (Note 6)	$ (58,953)	$ (71,488)
Investment in property and equipment, net (Note 4)	46,542	36,231
Contributions from foreign cooperative research agreements	3,997	3,632
Nuclear Waste Fund	15,024	19,592
Accounts receivable - fees	98,660	130,300
Fee collection revenue not transferred	(104)	-
Other	27	45
Cumulative Results of Operations	**$ 105,193**	**$ 118,312**

Note 9. STATEMENT OF NET COST

For the years ended September 30,	2011	2010
Nuclear Reactor Safety and Security		
Intragovernmental gross costs	$ 257,924	$ 272,871
Less: Intragovernmental earned revenue	(59,332)	(54,270)
Intragovernmental net costs	198,592	218,601
Gross costs with the public	599,644	609,720
Less: Earned revenues from the public	(727,408)	(782,033)
Net costs with the public	(127,764)	(172,313)
Total Net Cost of Nuclear Reactor Safety and Security	**$ 70,828**	**$ 46,288**
Nuclear Materials and Waste Safety and Security		
Intragovernmental gross costs	$ 71,987	$ 64,260
Less: Intragovernmental earned revenue	(7,686)	(7,314)
Intragovernmental net costs	64,301	56,946
Gross costs with the public	167,363	193,602
Less: Earned revenues from the public	(94,233)	(79,864)
Net costs with the public	73,130	113,738
Total Net Cost of Nuclear Materials and Waste Safety and Security	**$ 137,431**	**$ 170,684**

Chapter 3
FINANCIAL STATEMENTS AND AUDITOR'S REPORT

Note 10. EXCHANGE REVENUES

	2011	2010
Fees for licensing, inspection, and other services	$ 879,208	$ 912,794
Revenue from reimbursable work	9,452	10,687
Total Exchange Revenues	**$ 888,660**	**$ 923,481**

Note 11. FINANCING SOURCES OTHER THAN EXCHANGE REVENUE

	2011	2010
Appropriations Used		
Collections were used to reduce the fiscal year's appropriations recognized:		
Funds consumed	$1,060,178	$ 1,079,739
Less: Collection from fees assessed	(911,004)	(909,514)
Less: Nuclear Waste Funding expense	(14,548)	(33,112)
Total Appropriations Used	**$ 134,626**	**$ 137,113**

Funds consumed include $44.7 million and $81.1 million through
September 30, 2011, and 2010 respectively, of available funds from prior years.

	2011	2010
Non-Exchange Revenue		
Civil penalties	$ 98	$ 590
Miscellaneous receipts	172	879
Contra-Revenue	(270)	(1,469)
Total Non-Exchange Revenue	**$ -**	**$ -**

	2011	2010
Imputed Financing		
Civil Service Retirement System	$ 16,541	$ 19,895
Federal Employee Health Benefit	21,245	20,825
Federal Employee Group Life Insurance	92	92
Judgments/Awards	12,656	-
Total Imputed Financing	**$ 50,534**	**$ 40,812**

Note 12. TOTAL OBLIGATIONS INCURRED

	2011	2010
Direct Obligations		
Category A	$ 1,071,326	$ 1,079,158
Exempt from Apportionment	7,341	29,790
Total Direct Obligations	1,078,667	1,108,948
Reimbursable Obligations	4,818	10,137
Total Obligations Incurred	**$ 1,083,485**	**$ 1,119,085**

Obligations exempt from apportionment are the result of funds derived from the Nuclear Waste Fund. Category A Obligations consist of NRC appropriations only. Undelivered orders for the Nuclear Waste Fund are $5.0 million and $12.5 million, Salaries and Expenses $289.7 million and $288.1 million, and the Office of the Inspector General $2.2 million and $1.2 million through September 30, 2011, and 2010, respectively.

Note 13. NUCLEAR WASTE FUND

Included in NRC's budget for FY 2011 and 2010 are $9.9 million and $29.0 million, respectively, provided from the NWF. The Statement of Federal Financial Accounting Standards (SFFAS) No. 27, "Identifying and Reporting Earmarked Funds," lists three defining criteria for an earmarked fund. Generally, an earmarked fund is established by law to use specifically identified financing sources only for designated activities, and the statute provides explicit authority to retain current, unused revenues for future use. Also, the law includes a requirement to account for and report on the receipt and use of the financing sources as distinguished from general revenues.

Congress passed the *Nuclear Waste Policy Act of 1982* (Public Law 97-425) establishing the NWF to be administered by the DOE (42 U.S.C. 10222). Given the terms of the statute, the NWF clearly meets the definition of an earmarked fund from DOE's perspective, and DOE does indeed report the NWF as an earmarked fund in its Agency Financial Report.

For the NRC, the NWF transfer is a source of financing; its receipt of NWF funds is a use of NWF resources. The NRC collects no revenue on behalf of the NWF and has no administrative control over it. Furthermore, the Treasury has no separate fund symbol for the NWF under the NRC's agency location code. The receipt and expenditure of NWF money is reported to Treasury under the NRC's primary Salaries and Expenses fund (X0200).

Based on these facts, the NWF is not an earmarked fund from the NRC's perspective. In order to provide additional information to the users of these financial statements, enhanced disclosure of the fund is presented below.

The funding provided to the NRC in FY 2011 and FY 2010 was for the purpose of performing activities associated with DOE's application for a high-level waste repository at Yucca Mountain, NV. These activities included review of the application, conduct of thorough safety and security evaluations, preparation of the safety evaluation report, initiation of the inspection program, ensuring that the regulation process was made available to stakeholders and the general public, and providing legal advice and representation for staff reviews and Commission actions.

The NWF amounts received, expended, obligated, and unobligated balances as of September 30, 2011, and 2010 are shown in the following:

	2011	2010
Appropriations received	$ 9,980	$ 29,000
Expended appropriations	$ 14,601	$ 34,308
Obligations incurred	$ 7,341	$ 29,790
Unobligated balances	$ 9,996	$ 7,079

Note 14. EXPLANATION OF DIFFERENCES BETWEEN THE STATEMENT OF BUDGETARY RESOURCES AND THE BUDGET OF THE U.S. GOVERNMENT

The Statement of Federal Financial Accounting Standards (SFFAS) No. 7, "Accounting for Revenue and Other Financing Sources," requires the NRC to reconcile the budgetary resources reported on the Statement of Budgetary Resources to the prior fiscal year actual budgetary resources presented in the Budget of the U.S. Government and explain any material differences. The NRC does not have any material differences between the Statement of Budgetary Resources and the Budget of the U.S. Government.

Note 15. RECONCILIATION OF NET COST OF OPERATIONS TO BUDGETARY RESOURCES

For the years ended September 30,	2011	2010
Budgetary Resources Obligated		
Obligations incurred (Note 12)	$ 1,083,485	$ 1,119,085
Less: Spending authority from offsetting collections and recoveries	(33,334)	(33,799)
Less: Distributed offsetting receipts	(910,901)	(909,514)
Net Obligations	139,250	175,772
Other Resources		
Imputed financing from costs absorbed by others	50,534	40,812
Net Other Resources Used to Finance Activities	50,534	40,812
Total Resources Used to Finance Activities	189,784	216,584
Resources Used to Finance Items not Part of the Net Cost of Operations	(14,846)	(19,668)
Total Resources Used to Finance the Net Cost of Operations	174,938	196,916
Components of the Net Cost of Operations that will not require or generate resources in the current period	33,321	20,056
Net Cost of Operations	$ 208,259	$ 216,972

REQUIRED SUPPLEMENTARY INFORMATION
Schedule of Budgetary Resources (In Thousands)

For the fiscal year ended September 30, 2011	Salaries and Expenses	Office of Inspector General	Nuclear Facility Fees	Total
	X0200	X0300	X5280	
Budgetary Resources				
Unobligated balances, brought forward, October 1	$ 42,812	$ 1,887	$ -	$ 44,699
Recoveries of prior year obligations				
Actual	18,411	430	-	18,841
Budget authority				
Appropriation	1,043,463	10,860	(104)	1,054,219
Spending authority from offsetting collections				
Reimbursements earned-collected	12,400	39	-	12,439
Reimbursements earned-change in receivables	946	-	-	946
Change in unfilled customer orders-advance received	(3,506)	-	-	(3,506)
Change in unfilled customer orders-without advance	4,614	-	-	4,614
Subtotal-spending authority from offsetting collections	14,454	39	-	14,493
Permanently not available	(255)	(2)	-	(257)
Total Budgetary Resources	$ 1,118,885	$ 13,214	$ (104)	$ 1,131,995
Status of Budgetary Resources				
Obligations incurred (Note 12)				
Direct	$ 1,066,465	$ 12,202	$ -	$ 1,078,667
Reimbursable	4,818	-	-	4,818
Subtotal	1,071,283	12,202	-	1,083,485
Unobligated balance				
Apportioned	28,197	656	-	28,853
Exempt from apportionment	9,996	-	(104)	9,892
Subtotal	38,193	656	(104)	38,745
Unobligated balance, not available	9,409	356	-	9,765
Total Status of Budgetary Resources	$ 1,118,885	$ 13,214	$ (104)	$ 1,131,995
Change in Obligated Balance				
Obligated balance, net				
Unpaid obligations, brought forward, October 1	$ 374,425	$ 956	$ -	$ 375,381
Obligations incurred, net	1,071,283	12,202	-	1,083,485
Gross outlays	(1,076,362)	(12,034)	-	(1,088,396)
Recoveries of prior year obligations, actual	(18,411)	(430)	-	(18,841)
Change in uncollected customer payments, from Federal sources	(5,560)	-	-	(5,560)
Obligated balance, net, end of period				
Unpaid obligations	358,708	694	-	359,402
Uncollected customer payments, from Federal sources	(13,333)	-	-	(13,333)
Total unpaid obligated balance, net, end of period	$ 345,375	$ 694	$ -	$ 346,069
Net outlays				
Gross outlays	$ 1,076,362	$ 12,034	$ -	$ 1,088,396
Offsetting collections	(8,894)	(39)	-	(8,933)
Distributed offsetting receipts	-	-	(910,901)	(910,901)
Net Outlays	$ 1,067,468	$ 11,995	$ (910,901)	$ 168,562

Chapter 3
FINANCIAL STATEMENTS AND AUDITOR'S REPORT

SCHEDULE OF SPENDING (UNAUDITED)
(In Thousands)

For the fiscal year ended September 30, 2011		FY 2011
What money is available to spend?		
Total Resources	$	1,131,995
Less Amount Not Agreed to be Spent		(38,745)
Less Amount Not Available to be Spent		(9,765)
Total Amounts Agreed to be Spent	$	1,083,485
How was the money spent?		
Nuclear Reactor Safety and Security		
Payroll	$	503,200
Contracts		225,120
Travel		21,642
Rent, Communication and Utilities		34,614
Structures and Equipment		18,241
Other		48,309
Total Spending of Nuclear Reactor Safety and Security	$	851,126
Nuclear Materials and Waste Safety and Security		
Payroll	$	140,278
Contracts		62,757
Travel		6,033
Rent, Communication and Utilities		9,650
Structures and Equipment		5,079
Other		13,473
Total Spending of Nuclear Materials and Waste Safety and Security	$	237,270
Total Spending	$	1,088,396
Spending Related to Prior Year Amounts Agreed to be Spent		(4,911)
Total Amount Agreed to be Spent	$	1,083,485

SCHEDULE OF SPENDING (UNAUDITED)
(In Thousands)

For the fiscal year ended September 30, 2011	FY 2011
Who did the money go to?	
For Profit	$ 235,322
Individuals	593,521
Federal	241,100
State and Local Government	13,542
Total Amount Agreed to be Spent	**$ 1,083,485**
How was the money given?	
Payroll	$ 621,866
Contracts	297,755
Travel	26,976
Grants	15,357
Other	121,531
Total Amount Agreed to be Spent	**$ 1,083,485**

INSPECTOR GENERAL'S LETTER TRANSMITTING INDEPENDENT AUDITOR'S REPORT

**UNITED STATES
NUCLEAR REGULATORY COMMISSION**
WASHINGTON, D.C. 20555-0001

OFFICE OF THE
INSPECTOR GENERAL

November 9, 2011

MEMORANDUM TO: Chairman Jaczko

FROM: Hubert T. Bell /RA/
Inspector General

SUBJECT: RESULTS OF THE AUDIT OF THE UNITED STATES
NUCLEAR REGULATORY COMMISSION'S FINANCIAL
STATEMENTS FOR FISCAL YEAR 2011 (OIG-12-A-03)

The Chief Financial Officers Act of 1990, as amended (CFO Act), requires the Inspector General (IG) or an independent external auditor, as determined by the IG, to annually audit the United States Nuclear Regulatory Commission's (NRC) financial statements in accordance with applicable standards. In compliance with this requirement, the Office of the Inspector General (OIG) retained Urbach Kahn & Werlin, LLP, which merged with Clifton Gunderson, LLP (CG), to conduct this annual audit. Transmitted with this memorandum are the following CG reports:

- Opinion on the Principal Statements.

- Opinion on Internal Control.

- Compliance with Laws and Regulations.

NRC's Performance and Accountability Report includes comparative financial statements for FY 2011 and FY 2010. Therefore, it is important to note that Urbach Kahn & Werlin, LLP, performed the audit of NRC's FY 2010 financial statements. CG performed the audit of NRC's FY 2011 financial statements.

Objective of a Financial Statement Audit

The objective of a financial statement audit is to determine whether the audited entity's financial statements are free of material misstatement. An audit includes examining, on a test basis, evidence supporting the amounts and disclosures in the financial statements. An audit also includes assessing the accounting principles used and significant estimates made by management as well as evaluating the overall financial statement presentation.

CG's audit and examination were made in accordance with auditing standards generally accepted in the United States of America; *Government Auditing Standards* issued by the Comptroller General of the United States; attestation standards established by the American Institute of Certified Public Accountants; and Office of Management and Budget (OMB) Bulletin No. 07-04, *Audit Requirements for Federal Financial Statements*, as amended. The audit included, among other things, obtaining an understanding of NRC and its operations, including internal control over financial reporting; evaluating the design and operating effectiveness of internal control and assessing risk; and testing relevant internal controls over financial reporting. Because of inherent limitations in any internal control, misstatements due to error or fraud may occur and not be detected. Also, projections of any evaluation of the internal control to future periods are subject to the risk that the internal control may become inadequate because of changes in conditions, or that the degree of compliance with the policies or procedures may deteriorate.

FY 2011 Audit Results

The results are as follows:

 Financial Statements

- Unqualified opinion

 Internal Controls

- Unqualified opinion

 Compliance with Laws and Regulations

- No reportable instances of noncompliance/no substantial noncompliance noted

Office of the Inspector General Oversight of CG Performance

To fulfill our responsibilities under the CFO Act and related legislation for oversight of the quality of the audit work performed, we monitored CG's audit of NRC's FY 2011 financial statements by:

- Reviewing CG's audit approach and planning.

- Evaluating the qualifications and independence of CG's auditors.

- Monitoring audit progress at key points.

- Examining the working papers related to planning and performing the audit and assessing NRC's internal controls.

- Reviewing CG's audit reports for compliance with *Government Auditing Standards* and OMB Bulletin No. 07-04, as amended.

- Coordinating the issuance of the audit reports.

- Performing other procedures deemed necessary.

CG is responsible for the attached auditor's reports, dated November 7, 2011, and the conclusions expressed therein. OIG is responsible for technical and administrative oversight regarding the firm's performance under the terms of the contract. Our oversight, as differentiated from an audit in conformance with *Government Auditing Standards*, was not intended to enable us to express, and accordingly we do not express, an opinion on:

- NRC's financial statements.

- The effectiveness of NRC's internal control over financial reporting.

- NRC's compliance with laws and regulations.

However, our monitoring review, as described above, disclosed no instances where CG did not comply, in all material respects, with applicable auditing standards.

Meeting with the Chief Financial Officer

At the exit conference on November 8, 2011, representatives of the Office of the Chief Financial Officer, OIG, and CG discussed the results of the audit.

Comments of the Chief Financial Officer

In his response, the Chief Financial Officer (CFO) agreed with the report. The full text of the CFO's response follows this report.

We appreciate NRC staff's cooperation and continued interest in improving financial management within NRC.

Attachment: As stated

cc: Commissioner Svinicki
        Commissioner Apostolakis
        Commissioner Magwood
        Commissioner Ostendorff
        N. Mamish, OEDO
        K. Brock, OEDO
        C. Jaegers, OEDO

Clifton Gunderson LLP
Certified Public Accountants & Consultants

INDEPENDENT AUDITOR'S REPORT

Inspector General
United States Nuclear Regulatory Commission

Chairman
United States Nuclear Regulatory Commission

In our audit of the United States Nuclear Regulatory Commission (NRC) for fiscal year 2011, we found:

- The financial statements are presented fairly, in all material respects, in conformity with accounting principles generally accepted in the United States of America.
- The NRC maintained, in all material respects, effective internal control over financial reporting.
- We noted no reportable instances of noncompliance with federal financial management systems requirements, applicable Federal accounting standards, and the United States Government Standard General Ledger (USSGL) at the transaction level.
- We noted no reportable instances of noncompliance with laws and regulations we tested.

The following sections discuss in more detail (1) these conclusions and our conclusions relating to other information presented in the Management's Discussion and Analysis and other supplementary information, (2) management's responsibilities, and (3) our objectives, scope and methodology.

Opinion on the Financial Statements

In our opinion, the financial statements including the accompanying notes present fairly, in all material respects, the financial position of the NRC as of September 30, 2011 and 2010, and its net cost, changes in net position, and budgetary resources for the years then ended, in conformity with accounting principles generally accepted in the United States of America. The financial statements of NRC as of September 30, 2010 were audited by Urbach Kahn & Werlin LLP, which practice was acquired by Clifton Gunderson LLP by merger on March 22, 2010. Urbach Kahn & Werlin LLP's report dated November 7, 2010, expressed an unqualified opinion on those financial statements.

4250 N. Fairfax Drive, Suite 1020
Arlington, Virginia 22203
tel: 571.227.9500
fax: 571.227.9552
www.cliftoncpa.com   Offices in 17 states and Washington DC

HLB International

INDEPENDENT AUDITOR'S REPORT, CONTINUED

Opinion on Internal Control

In our opinion, the NRC maintained, in all material respects, effective control over financial reporting as of September 30, 2011, that provided reasonable assurance that misstatements, losses or noncompliance material in relation to the financial statements would be prevented, or detected and corrected, on a timely basis. Our opinion is based on criteria established under 31 U.S.C. 3512 (c), (d), the Federal Managers' Financial Integrity Act (FMFIA), and the Office of Management and Budget (OMB) Circular No. A-123, *Management's Responsibility for Internal Control*, and Government Accountability Office's (GAO's) *Standards for Internal Control in the Federal Government*, as required by OMB Bulletin 07-04, *Audit Requirements for Federal Financial Statements*, as amended.

Compliance with Laws and Regulations

Under the Federal Financial Management Improvement Act (FFMIA), we are required to report whether the NRC's financial management systems substantially comply with federal financial management systems requirements, applicable Federal accounting standards, and the USSGL at the transaction level. To meet this requirement, we performed tests of compliance with the provisions of FFMIA Section 803(a). The results of our tests disclosed no substantial noncompliance with federal financial management systems requirements, applicable Federal accounting standards, and the USSGL at the transaction level.

The results of our tests of compliance with laws and regulations disclosed no instances of noncompliance that are required to be reported under *Government Auditing Standards* and OMB Bulletin No. 07-04, *Audit Requirements for Federal Financial Statements*, as amended. Providing an opinion on compliance with laws and regulations was not an objective of our audit and, accordingly, we do not express such an opinion.

Other Information

The information in Management's Discussion and Analysis and other Required Supplementary Information in NRC's Performance and Accountability Report is not a required part of the financial statements, but is supplementary information required by accounting principles generally accepted in the United States of America. We have applied certain limited procedures, which consisted principally of inquiries of management regarding the methods of measurement and presentation of the supplementary information. However, we did not audit the information and express no opinion on it.

The Program Performance and Other Accompanying Information sections listed in the Table of Contents are presented for additional analysis and are not a required part of the financial statements. Such information has not been subjected to the auditing procedures applied in the audit of the financial statements and, accordingly, we express no opinion on them.

INDEPENDENT AUDITOR'S REPORT, CONTINUED

Management Responsibilities

Management is responsible for (1) preparing the financial statements in conformity with accounting principles generally accepted in the United States of America, (2) establishing and maintaining effective internal control over financial reporting, and evaluating its effectiveness, (3) ensuring that the NRC's financial management systems substantially comply with FFMIA requirements, and (4) complying with applicable laws and regulations. NRC management evaluated the effectiveness of NRC's internal control over financial reporting as of September 30, 2011, based on criteria established under FMFIA. NRC management's assurances are included in the Management's Discussion and Analysis.

Objectives, Scope and Methodology

We are responsible for planning and performing our audit to obtain reasonable assurance about whether the financial statements are free of material misstatement. An audit includes examining, on a test basis, evidence supporting the amounts and disclosures in the financial statements. An audit also includes assessing the accounting principles used and significant estimates made by management, as well as evaluating the overall financial statement presentation.

We are responsible for planning and performing our examination to obtain reasonable assurance about whether management maintained, in all material respects, effective internal control over financial reporting as of September 30, 2011. Our examination included obtaining an understanding of NRC and its operations, including internal control over financial reporting; considering NRC's process for evaluating and reporting on internal control over financial reporting which the NRC is required to perform by FMFIA; assessing the risk that a material misstatement exists in the financial statements and the risk that a material weakness exists in internal control over financial reporting; evaluating the design and operating effectiveness of internal control and assessing risk; testing relevant internal controls over financial reporting; and performing such other procedures as we considered necessary in the circumstances. We did not test all internal controls relevant to operating objectives as broadly defined by FMFIA.

An entity's internal control over financial reporting is a process effected by those charged with governance, management, and other personnel, the objectives of which are to provide reasonable assurance that (1) transactions are properly recorded, processed, and summarized to permit the preparation of financial statements in accordance with accounting principles generally accepted in the United States of America, and assets are safeguarded against loss from unauthorized acquisition, use, or disposition; and (2) transactions are executed in accordance with the laws governing the use of budget authority and other laws and regulations that could have a direct and material effect on the financial statements.

Because of inherent limitations in any internal control, misstatements due to error or fraud may occur and not be detected. Also, projections of any evaluation of the internal control to future periods are subject to the risk that the internal control may become inadequate because of changes in conditions, or that the degree of compliance with the policies or procedures may deteriorate.

INDEPENDENT AUDITOR'S REPORT, CONTINUED

We are also responsible for testing compliance with selected provisions of laws and regulations that have a direct and material effect on the financial statements. We did not test compliance with all laws and regulations applicable to the NRC. We limited our tests of compliance to selected provisions of laws and regulations that have a direct and material effect on the financial statements and those required by OMB audit guidance that we deemed applicable to the financial statements for the fiscal year ended September 30, 2011. We caution that noncompliance may occur and not be detected by these tests and that such testing may not be sufficient for other purposes.

We conducted our audit and examinations in accordance with auditing standards generally accepted in the United States of America, *Government Auditing Standards*, issued by the Comptroller General of the United States; attestation standards established by the American Institute of Certified Public Accountants; and OMB Bulletin No. 07-04, *Audit Requirements for Federal Financial Statements*, as amended. We believe that our audit and examinations provide a reasonable basis for our opinions.

We noted less significant matters involving the NRC's internal control and its operation, which we have reported to agency management separately.

Distribution

This report is intended solely for the information and use of the NRC Office of Inspector General, the management of NRC, OMB, the GAO, and the United States Congress, and is not intended to be and should not be used by anyone other than these specified parties.

Clifton Gunderson LLP

Arlington, Virginia
November 7, 2011

MANAGEMENT'S RESPONSE TO THE INDEPENDENT AUDITOR'S REPORT ON THE FINANCIAL STATEMENTS

UNITED STATES
NUCLEAR REGULATORY COMMISSION
WASHINGTON, D.C. 20555-0001

OFFICE OF THE
CHIEF FINANCIAL OFFICER

November 9, 2011

MEMORANDUM TO: Stephen D. Dingbaum
Assistant Inspector General for Audits
Office of the Inspector General

FROM: J. E. Dyer
Chief Financial Officer

SUBJECT: AUDIT OF THE FISCAL YEAR 2011 AND 2010 FINANCIAL STATEMENTS

We appreciate the collaborative relationship between the Office of the Inspector General, the auditors, and the Office of the Chief Financial Officer in supporting our continuing effort to improve financial reporting. We have reviewed the Independent Auditor's Report of the Agency's Fiscal Year 2011 and 2010 financial statements and are in agreement with it.

cc: N. Mamish, AO/OEDO
J. Arildsen, OEDO
C. Jaegers, OEDO
D. Holley, OCFO

Chapter 4

OTHER ACCOMPANYING INFORMATION

UNITED STATES
NUCLEAR REGULATORY COMMISSION
WASHINGTON, D.C. 20555-0001

OFFICE OF THE
INSPECTOR GENERAL

October 3, 2011

MEMORANDUM TO: Chairman Jaczko

FROM: Hubert T. Bell /RA/
                    Inspector General

SUBJECT: INSPECTOR GENERAL'S ASSESSMENT
                    OF THE MOST SERIOUS MANAGEMENT
                    AND PERFORMANCE CHALLENGES
                    FACING NRC (OIG-12-A-01)

The *Reports Consolidation Act of 2000* requires the Inspector General of each Federal agency to annually summarize what he or she considers to be the most serious management and performance challenges facing the agency and to assess the agency's progress in addressing those challenges. In accordance with the act, I identified seven management and performance challenges confronting the Nuclear Regulatory Commission that I consider to be the most serious.

We appreciate the cooperation extended to us during this evaluation. The agency provided comments on this report, which have been incorporated as appropriate. If you have any questions, please contact Stephen D. Dingbaum, Assistant Inspector General for Audits, at 415-5915 or me at 415-5930.

Attachment: As stated

Inspector General's Assessment of the Most Serious Management and Performance Challenges Facing NRC

EXECUTIVE SUMMARY

BACKGROUND

The *Reports Consolidation Act of 2000* requires the Inspector General (IG) of each Federal agency to annually summarize what he or she considers to be the most serious management and performance challenges facing the agency and to assess the agency's progress in addressing those challenges.

OBJECTIVE

In accordance with the act, the IG at the U.S. Nuclear Regulatory Commission (NRC) updated what he considers to be the most serious management and performance challenges facing NRC. The IG considered the overall work of the Office of the Inspector General (OIG), the OIG staff's general knowledge of agency operations, and other relevant information to develop and update his list of management and performance challenges. In addition, OIG staff sought input from NRC's Chairman, Commissioners, and management to obtain their views on what challenges the agency is facing and what efforts the agency has taken or are underway or planned to address previously identified management and performance challenges.

RESULTS IN BRIEF

The IG identified seven challenges that he considers the most serious management and performance challenges facing NRC. The challenges identify critical areas or difficult tasks that warrant high-level management attention.

The 2011 list of challenges reflects two changes from the 2010 list. Prior Challenge 1, *Protection of nuclear material used for civilian purposes*, was reworded to *Oversight of nuclear material used for civilian purposes*. This change was made to more accurately describe NRC's regulatory oversight role relative to nuclear material as NRC does not directly protect nuclear material, but provides oversight of licensees who are charged to protect the material. Prior Challenge 3, *Ability to modify regulatory processes to meet a changing environment, to include the licensing of new nuclear facilities*, was reworded to reflect changing economic conditions for new facility construction, as well as ongoing efforts to evaluate post-Fukushima Dai-ichi lessons learned for NRC's oversight of currently operating facilities. Current Challenge now 3 reads *Ability to modify regulatory processes to meet a changing environment in the oversight of nuclear facilities.*

The following chart provides an overview of the seven most serious management and performance challenges as of October 1, 2011.

Most Serious Management and Performance Challenges Facing the Nuclear Regulatory Commission as of October 1, 2011* *(as identified by the Inspector General)*	
Challenge 1	*Oversight of nuclear material used for civilian purposes.*
Challenge 2	*Managing information to balance security with openness and accountability.*
Challenge 3	*Ability to modify regulatory processes to meet a changing environment in the oversight of nuclear facilities.*
Challenge 4	*Oversight of radiological waste.*
Challenge 5	*Implementation of information technology and information security measures.*
Challenge 6	*Administration of all aspects of financial management and procurement.*
Challenge 7	*Managing human capital.*

The most serious management and performance challenges are not ranked in any order of importance.

Inspector General's Assessment of the Most Serious Management and Performance Challenges Facing NRC

CONCLUSION

The seven challenges contained in this report are distinct, yet interdependent relative to the accomplishment of NRC's mission. For example, the challenge of managing human capital affects all other management and performance challenges.

The agency's continued progress in taking actions to address the challenges presented should facilitate achieving the agency's mission and goals.

ABBREVIATIONS AND ACRONYMS

ADSA	Associate Directorate for Strategic Acquisition
ASLB	Atomic Safety Licensing Board
CUI	controlled unclassified information
DOE	U.S. Department of Energy
FAIMIS	Financial Accounting and Integrated Management Information System
GALL	Generic Aging Lessons Learned
IG	Inspector General
ITAAC	inspections, tests, analyses, and acceptance criteria
NRC	U.S. Nuclear Regulatory Commission
OIG	Office of the Inspector General
3WFN	Three White Flint North

TABLE OF CONTENTS

I.   BACKGROUND

On January 24, 2000, Congress enacted the *Reports Consolidation Act of 2000* (Reports Act), requiring Federal agencies to provide financial and performance management information in a more meaningful and useful format for Congress, the President, and the public. The Reports Act requires the Inspector General (IG) of each Federal agency to annually summarize what he or she considers to be the most serious management and performance challenges facing the agency and to assess the agency's progress in addressing those challenges.

II.   OBJECTIVE

In accordance with the Reports Act's provisions, the U.S. Nuclear Regulatory Commission (NRC) IG updated what he considers to be the most serious management and performance challenges facing the agency. The IG considered the overall work of the Office of the Inspector General (OIG), the OIG staff's general knowledge of agency operations, and other relevant information to develop and update his list of management and performance challenges.

In addition, OIG staff sought input from NRC's Chairman, Commissioners, and management to obtain their views on what challenges the agency is facing and what efforts the agency has taken or are underway or planned to address previously identified management and performance challenges.

Inspector General's Assessment of the Most Serious Management and Performance Challenges Facing NRC

III.   EVALUATION RESULTS

The NRC's mission is to license and regulate the Nation's civilian use of byproduct, source, and special nuclear materials to ensure adequate protection of public health and safety, promote the common defense and security, and protect the environment. Like other Federal agencies, NRC faces management and performance challenges in carrying out its mission.

Determination of Management and Performance Challenges

Congress left the determination and threshold of what constitutes a most serious management and performance challenge to the discretion of the IGs. As a result, the IG applied the following definition in identifying challenges:

Serious management and performance challenges are mission critical areas or programs that have the potential for a perennial weakness or vulnerability that, without substantial management attention, would seriously impact agency operations or strategic goals.

Based on this definition, in 2011, the IG assessed the most serious management and performance challenges facing NRC and identified seven challenges that he considered most serious. The challenges identify critical areas or difficult tasks that warrant high-level management attention. The 2011 list of challenges reflects two changes from the 2010 list:

- Prior Challenge 1, *Protection of nuclear material used for civilian purposes*, was reworded to *Oversight of nuclear material used for civilian purposes*. This change was made to more accurately describe NRC's regulatory oversight role relative to nuclear material as NRC does not directly protect nuclear material, but provides oversight of licensees who are charged to protect the material.

- Prior Challenge 3, *Ability to modify regulatory processes to meet a changing environment, to include the licensing of new nuclear facilities*, was reworded to reflect changing economic conditions for new facility construction, as well as ongoing efforts to evaluate post-Fukushima Dai-ichi lessons learned for NRC's oversight of currently operating facilities. Current Challenge 3 now reads *Ability to modify regulatory processes to meet a changing environment in the oversight of nuclear facilities*.

The following chart provides an overview of the seven challenges identified as most serious. The sections that follow the chart provide more detailed descriptions of the challenges, descriptive examples related to the challenges, and examples of efforts that the agency has taken or are underway or planned to address the challenges.

Inspector General's Assessment of the Most Serious Management and Performance Challenges Facing NRC

Most Serious Management and Performance Challenges Facing the Nuclear Regulatory Commission as of October 1, 2011*
(as identified by the Inspector General)

Challenge 1	*Oversight of nuclear material used for civilian purposes.*
Challenge 2	*Managing information to balance security with openness and accountability.*
Challenge 3	*Ability to modify regulatory processes to meet a changing environment in the oversight of nuclear facilities.*
Challenge 4	*Oversight of radiological waste.*
Challenge 5	*Implementation of information technology and information security measures.*
Challenge 6	*Administration of all aspects of financial management and procurement.*
Challenge 7	*Managing human capital.*

*The most serious management and performance challenges are not ranked in any order of importance.

Inspector General's Assessment of the Most Serious Management and Performance Challenges Facing NRC

CHALLENGE 1
Oversight of nuclear material used for civilian purposes.

NRC is authorized to grant licenses for the possession and use of radioactive materials and establish regulations to govern the possession and use of those materials.

NRC's regulations require that certain material licensees have extensive material control and accounting programs as a condition of their licenses. All other license applicants (including those requesting authorization to possess small quantities of special nuclear materials) must develop and implement plans that demonstrate a commitment to accurately control and account for radioactive materials.

NRC may relinquish to States, upon their request, its authority to regulate certain radioactive materials and limited quantities of special nuclear material. After these States demonstrate that their regulatory programs are adequate to protect public health and safety and compatible with NRC's program, the States enter into an agreement assuming this regulatory authority from NRC and are called Agreement States.

The issues related to this challenge and the agency's actions to address each issue include the following:

Issue: Implement the National Source Tracking System, Web Based Licensing, and the Licensing Verification System to ensure the accurate tracking and control of byproduct material, especially those  materials with the greatest potential to impact public health and safety.

NSTS logo. Source: NRC

Action: Since the National Source Tracking System became operational in December 2008, NRC has continued implementation of and improvements to the system. Version 2.0 was deployed in May 2011. Revisions included functionality improvements designed to broaden system capabilities for all users. NRC is also continuing development of Web-Based Licensing and the License Verification System. Further, NRC is working to integrate these systems with the National Source Tracking System to license and track source materials under one management mechanism.

Issue: Ensure that radioactive material is adequately protected to preclude its use for malicious purposes.

Code of Conduct.
Source:
www.iaea.org

Action: Although NRC initiated a rulemaking to expand the materials tracked in the National Source Tracking System, the decision and potential implementation of that rulemaking was not approved by the Commission. As a result, the system is available to licensees to report transactions involving Code of Conduct[1] materials in only categories 1 and 2. NRC has provided licensees with multiple ways of reporting such transactions. About 60 percent of transactions are submitted electronically and about 40 percent are submitted via facsimile machine, U.S. Postal Service, or e-mail. NRC continues to work on getting more licensees to use the online system and contracted with a local marketing firm to help improve the online use of the system.

In addition to collecting information in the National Source Tracking System, NRC launched a pilot inspection program to collect information on self-shielded irradiators and other irradiators. The purpose of the pilot is to assess the need to modify the current inspection program to determine if more frequent inspections would result in greater compliance with security requirements. One radioactive compound used in irradiators that is of particular concern for malicious use is cesium-chloride. The U.S. National Academy of Sciences issued a report emphasizing that replacement technologies be considered for cesium-chloride, a highly dispersible chemical form of the radioactive isotope of Cesium, Cs-137. Cesium-chloride is very soluble in water and

---

[1] In January 2004, the International Atomic Energy Agency published the Code of Conduct on the Safety and Security of Radioactive Sources as the standard the international community uses to govern the safety and security of radioactive materials based on the categorization system. While the International Atomic Energy Agency classifies sources into five categories, it notes that sources in categories one through three are designated as varying degrees of dangerous.

easily dispersed in the air and is highly toxic if ingested. Cesium-chloride, used in nuclear medicine, research, and industry, is typically double sealed and contained in a stainless steel capsule for safety reasons. In light of the views on alternative technologies as a replacement, NRC convened public workshops to seek input from various stakeholders. NRC also commissioned a study by its *Advisory Committee on the Medical Uses of Isotopes*. After carefully considering all these inputs, as well as the NRC's own internal analysis, the agency concluded that near-term replacement of cesium-chloride devices was not practicable, and would be detrimental to the delivery of medical care and research. As a result, NRC updated its policy statement, which still allows for the safe and secure use of cesium chloride and further states that the development and use of alternative forms of the material are prudent but not required.

Issue: Ensure the appropriate oversight of uranium recovery facilities.

Action: NRC maintains a regulatory oversight program with respect to licensing and inspection of uranium recovery facilities to ensure that licensees conduct activities safely and in an environmentally protective manner. NRC regulates six in situ[2] recovery facilities,[3] one conventional mill, and 11 mill sites undergoing decommissioning in the Western States.

Crow Butte In Situ Recovery Facility. Source: NRC Web site.

Additionally, NRC works with the Department of Energy (DOE) regarding abandoned uranium mill tailing sites that are covered by Title I of the Uranium Mill tailings Radiation Control Act of 1978.

---

[2] In situ recovery is one of the two primary extraction methods that are currently used to obtain uranium from underground. These facilities recover uranium from low-grade ores where other mining and milling methods may be too expensive or environmentally disruptive.
[3] One of the five in situ recovery facilities has been in litigation for 10 years and there has been no construction at that site; therefore, NRC's regulatory oversight is minimal.

Under a general license from NRC, DOE is responsible for cleanup and remediation of these sites.

Issue: Ensure reliable accounting of special nuclear materials in the NRC and DOE jointly managed Nuclear Materials Management and Safeguards System.

Action: NRC has been working for over 8 years to resolve issues of material control and accounting in response to OIG's 2003 report, OIG-03-A-15, *Audit of NRC's Regulatory Oversight of Special Nuclear Materials*. On February 7, 2008, NRC approved a final rule that amended its regulations to improve the accuracy of material inventory information maintained in the Nuclear Materials Management and Safeguards System. The amendments, effective January 1, 2009, lower the threshold of reportable quantities of special nuclear materials and certain source materials to the Nuclear Materials Management and Safeguards System, modify the types and timing of submittals to the system, and require licensees to reconcile any material inventory discrepancies that NRC identifies in the system database. NRC reports that it has implemented the rule change requiring improved reporting and reconciliation for licensees reporting to the Nuclear Materials Management and Safeguards System, and continues to verify the adequacy of material control and accounting of special nuclear material at NRC licensed facilities. Additionally, the Commission has directed the NRC staff to revise and consolidate current material control and accounting regulations into Title 10, Code of Federal Regulations, Part 74. The proposed rule is scheduled to be published for public comment in late 2011. The final rule and associated guidance is scheduled to be completed by August 30, 2012.

Issue: Ensure that Agreement State programs are adequate to protect public health and safety and the environment, and are compatible with NRC's program.

Action: NRC conducts 8 to 10 reviews per year of Agreement State radioactive materials programs and NRC's regional programs under the agency's *Integrated Materials Performance Evaluation Program*. Furthermore, NRC completed a self-assessment of the program in July 2010. NRC management endorsed the self-assessment report and the agency initiated actions to address the report's 15 recommendations.

Agreement and Non-Agreement States.
Source: NRC Web site

Issue: Ensure appropriate oversight of nuclear materials used in medicine.

Action: The agency is holding a series of public workshops to address concerns regarding the continued disagreement in the medical community about the correct approach for a definition of a medical event. During the workshops, NRC will solicit stakeholder input on topics associated with the medical event definition, including sections involving reporting and notifications of medical events for permanent implant brachytherapy and other medical issues that are currently being considered for rulemaking.

NRC is engaged in several activities to address concerns about doses given to members of the public from patients who have been treated with radioactive isotopes. For example, the agency expanded its guidance requirement for doctors to include advice that patients are strongly discouraged from checking into hotels immediately following treatment. Additionally, NRC staff are examining whether there are gaps in the available data regarding doses being received by members of the public due to the release of patients treated with medical isotopes, as well as how the agency could go about collecting additional data, if needed.

CHALLENGE 2
Managing information to balance security with openness and accountability.

NRC employees create and work with a significant amount of sensitive information that needs to be protected. Such information includes sensitive unclassified information and classified national security information contained in written documents and various electronic databases.

Based on continuing national security threats, NRC continually reexamines its information management policies and procedures. NRC faces the challenge of attempting to balance the need to protect sensitive information from inappropriate disclosure with the agency's goal of openness in its regulatory processes. Over the past year, NRC has made various efforts to improve public access to information while protecting sensitive information, including security-related information, from inappropriate disclosure.

The issues related to this challenge and the agency's actions to address each issue include the following:

Issue: Respond to requests for information and provide external stakeholders with clear and accurate information about regulatory programs and facilitate public participation in the regulatory process to ensure openness and accountability.

> **Action:** NRC has published datasets of regulatory information (e.g., inspection reports, event reports) on the Data.Gov Web site per the Administration's Open Government initiative. This initiative is intended to increase public knowledge and promote transparency by providing the public with access to machine-readable Government information for use in research and analysis.

Inspector General's Assessment of the Most Serious Management and Performance Challenges Facing NRC

Issue: Manage information in accordance with new Federal Government policies for designating, marking, safeguarding, and disseminating controlled unclassified information (CUI).

Action: The agency participated in working groups with the National Archives and Records Administration's CUI Office to develop a CUI Executive Order and Implementing Directive. NRC has submitted its catalog of proposed CUI categories, subcategories, and markings to the National Archives and Records Administration.

Issue: Ensure that sensitive information is handled in accordance with agency policies and procedures for public disclosure.

Action: NRC ensured that Privacy Act compliance activities were completed, such as Privacy Impact Assessments, requirements of the Office of Management and Budget Circular A-130, and Office of Management and Budget requirements for personal identifiable information.

Action: In response to recommendations in OIG's *Audit of the Shared S Drive* (OIG-11-A-15), the agency will revise its Personally Identifiable Information and information security training, provide agency information technology coordinators with role-based training, develop CUI policies and guidance for storing and protecting CUI in agency shared drives, and implement procedures for quality assurance checks following network upgrades to ensure that access controls are preserved in shared network drives that process documents containing CUI.

Issue: Review and strengthen programs to protect licensee, vendor, and Government-owned assets (e.g., facility designs, technology descriptions, dual use material and components, classified information) from compromise by foreign sources and industrial espionage and increase awareness of the relationship of these assets to the Nation's economic and industrial base and energy infrastructure.

Inspector General's Assessment of the Most Serious Management and Performance Challenges Facing NRC

Action: NRC has recognized the need to ensure technological data involving licensee, vendor, and Government-owned assets is fully protected against potential loss to adversaries. NRC has promulgated orders that provide additional security measures for the protection of these assets.

NRC employees and contractors are required to have a baseline level of security awareness upon entry on duty and a security clearance. Some, depending on their job and involvement in the creation and use of protected information, are provided various additional "role based" training programs, such as classifier's training, training for administrative personnel, declassification training, Secret Internet Protocol Router Network users training, and Sensitive Compartmented Information access training. The training is layered, targeted, and recurring for those with specific responsibilities for various types of protected information.

In addition, NRC has increased its information security awareness by issuing various agencywide announcements informing staff of methods used to target NRC information systems and the corresponding need for employees to heighten their computer security information protection posture.

Issue: Technologies and materials, which NRC regulates, have potential intelligence value to foreign states and non-state actors from either an intelligence or a counterproliferation, counterterrorism, or economic espionage perspective and should be protected from potential compromise. Further, there is the potential that NRC employees have knowledge and access to information that may be of interest to foreign powers and non-state actors.

Action: NRC has begun the process of developing programmatic efforts aimed at identifying potential threats and vulnerabilities that exist in its programs and operations. Such efforts should continue and receive senior leadership support.

CHALLENGE 3
Ability to modify regulatory processes to meet a changing environment in the oversight of nuclear facilities.

NRC faces the challenge of maintaining its core regulatory programs while adapting to changes in its regulatory environment. NRC must address a steady interest in licensing and constructing new nuclear power plants to meet the Nation's increasing demands for energy production. As of May 2011, NRC had received 18 Combined Operating License applications, 12 of which NRC was actively reviewing. Moreover, the agency is reviewing three standard design certifications and two certification amendments, and expects to receive five new advanced design certification applications through 2013.

While responding to the emerging demands associated with licensing and regulating new reactors, NRC must maintain focus and effectively carry out its current regulatory responsibilities, such as inspections of the current fleet of operating nuclear reactors and fuel cycle facilities. NRC intends to increase its safety focus on licensing and oversight activities through risk-informed and performance-based regulation.

The issues related to this challenge and the agency's actions to address each issue include the following:

New Facilities

Issue: Implement the new Construction Inspection Program.

- Risk-inform Construction Inspection Program activities to ensure the safe operation of newly constructed nuclear facilities.

- Ensure that the NRC staff has the necessary knowledge and skill to successfully implement the program.

    Action: NRC has developed a Construction Inspection Program in accordance with Title 10, Code of Federal Regulations, Part 52. New inspections, tests, analyses, and acceptance criteria (ITAAC) have been integrated into the Part 52 licensing process to create a "design-specific, pre-approved set of performance standards that the licensee must meet to the NRC's satisfaction." While the agency is still developing ITAAC review processes and procedures, it has conducted an ITAAC demonstration project to simulate and test numerous aspects of the ITAAC inspection, closure, and verification process. The agency also created the Center for

Construction Inspection in Region II. The Center's mission is to provide assurance in the safety of future operations at new nuclear facilities by ensuring that licensees and applicants construct the facilities according to approved design criteria, using appropriate practices and quality materials.

NRC Conducts Inspections.
Source: NRC Web site

NRC continues to make improvements to its construction inspection and quality assurance practices consistent with OIG recommendations. For example, during Fiscal Year 2011, the agency completed revisions to NRO-REG-112, "New Reactor Construction Experience Program," to improve the screening of construction operating experience for the purposes of documenting lessons learned.

Issue: As the public's demand for new energy sources continues, NRC must ensure that the process for reviewing applications for new facilities focuses on safety and effectiveness.

Action: NRC's review of Combined Operating License applications has been complicated because some applicants are revising the reactor design currently under review. The agency staff is working with applicants to resolve issues related to design, siting, and schedule modifications using a variety of interactions. These include pre-application audits, site visits, reviews of topical/technical reports, public meetings, and Design-Centered Working Group meetings.

NRC is also undertaking pre-application interactions with vendors of advanced and small modular reactor designs. To facilitate the licensing of new reactor designs that differ from the current generation of large light water reactor facilities, the NRC staff seeks to resolve key safety and licensing issues and develop a regulatory infrastructure to support licensing review of these unique reactor designs. The staff has identified several potential policy and technical issues associated with licensing small light water reactor and non-light water reactor designs.

Issue: As the sources of manufactured reactor components become more globalized, NRC must ensure its regulations and oversight activities appropriately address the challenges associated with licensees procuring components from suppliers located outside of the United States.

Action: NRC continues to make improvements to its vendor oversight activities consistent with OIG recommendations. In late December 2010, the NRC staff began developing an agencywide approach to address the threat of counterfeit and fraudulent components. Furthermore, in Fiscal Year 2011, NRC updated vendor inspection procedures to establish expectations that translators and interpreters will be used as necessary to ensure that the use of foreign-language documents or communication with foreign-language speakers does not degrade the quality of the inspection. NRC also participates in the Multinational Design Evaluation Program, which is a multinational initiative taken by national safety authorities to develop approaches to leverage the resources and knowledge of the national regulatory authorities who will be tasked with the review of new reactor power plant designs.

Existing Fleet

Issue: Ensure NRC maintains the ability to effectively review licensee applications for license renewals and power uprates submitted by industry in response to the Nation's increasing demands for energy production.

Action: For planning purposes, NRC continues to work with plant licensees to develop a schedule of anticipated license amendment requests for license renewals and power uprates. The agency has also implemented a number of recommendations to improve the license renewal review and power uprate processes to include closer management oversight of report-writing standards and technical reviewer and project manager training.

As NRC has gained experience with Inspection Procedure 71003, "Post-Approval Site Inspection for License Renewal," during recent years, the staff provided an updated "Frequently Asked Questions" document on the NRC Web site in March 2011. Furthermore, the staff updated a major license renewal document, NUREG-1801, "Generic Aging Lessons Learned (GALL) Report," in December 2010.

Issue: Respond to a heightened public focus on license renewals resulting in contested hearings.

Action: NRC has open dialogs with the industry, licensees, and stakeholders. The public, in general, is also encouraged to participate in the process through public meetings, and public comment periods on rules, renewal guidance, and other documents. For example, NRC routinely requests public comment on supplements of NUREG-1437, "License Renewal Generic Environmental Impact Statement." In addition, the public has an opportunity to request a formal adjudicatory hearing if that party would be adversely affected by the renewal.

Issue: Ensure the ability to identify emerging operating and safety issues at all plants, including issues associated with license renewal and power uprate; consistently apply regulatory and review changes in response to these emerging issues across the existing fleet of reactors.

Action: NRC continues to evaluate the need to make changes to its regulatory programs based on emerging operational and safety issues related to license renewal and power uprate. For example, after the June 2010 Groundwater Contamination Task Force report was issued with recommendations for the agency to strengthen its response to groundwater incidents, the staff initiated, and plans to continue, efforts to work with consensus standards organizations to have certain provisions related to inspecting and maintaining buried piping incorporated into applicable industry standards.

Cover from NRC's Post-Fukushima Accident Safety Report.
Source: NRC Web site

In March 2011, after an earthquake and tsunami struck Japan affecting several reactors at the Fukushima Dai-ichi site, NRC activated its 24-hour Emergency Operations Center to monitor and analyze events at the nuclear plants in Japan. Soon thereafter, NRC established the senior-level "Near-Term Task Force Review of Insights from the Fukushima Dai-ichi Accident" to conduct a methodical and systematic review of the NRC's processes and regulations to determine whether the agency should make additional improvements to its regulatory system and to make recommendations to the Commission for its policy direction. In 2011, the Task Force delivered a report to the Commission with 12 recommendations for both short- and long-term changes to NRC's oversight of nuclear reactors. These recommendations focused on issues such as loss of electrical power, spent fuel pools, natural disaster preparedness and recovery, and NRC's overall regulatory approach to oversight. NRC's next steps are to determine how best to implement the Task Force recommendations, continue to evaluate new information coming from Fukushima Dai-ichi, and identify, review, and address lessons learned regarding the agency's and the Federal Government's response regarding the

Fukushima Dai-ichi event, and balance these emerging efforts against ongoing work.

Issue: Establish and maintain effective, stable, and predictable regulatory programs or policies for all programs.

Action: NRC continues to interface with stakeholders, develop regulatory policy, update rules and technical guidance, provide technical lead and management for the Reactor Oversight Process, and support the development of programmatic changes when needed. Additionally, the Reactor Oversight Process features an annual assessment process which is used to revise the program as necessary.

CHALLENGE 4
Oversight of radiological waste.

NRC regulates spent nuclear fuel generated from commercial nuclear power reactors, referred to as high-level radioactive waste. NRC faces significant issues involving the uncertainty of Yucca Mountain as the Nation's repository for storing high-level radioactive waste. Other challenges in the high-level waste area include the interim storage of spent nuclear fuel, certification of storage and transportation casks, and the oversight of decommissioned reactors and other nuclear sites.

Additionally, the amount of low-level waste continues to grow; however, no new disposal facilities have been built since the 1980s, and unresolved issues will increase as access to disposal facilities becomes more limited given facility closures and restricted accessibility.

The issues related to this challenge and the agency's actions to address each issue include the following:

Issue: Address issues regarding DOE's March 3, 2010, motion to withdraw its license application to construct a high-level radioactive waste repository at Yucca Mountain.

Action: NRC stopped its review of DOE's high-level radioactive waste repository application despite a June 2010 determination by the NRC's Atomic Safety and Licensing Board (ASLB) that denied DOE's motion to withdraw the application. The ASLB grounded its decision in its interpretation of the Nuclear Waste Policy Act, reasoning that Congress directed DOE to file the application and NRC to consider the application and issue a decision based on its technical merits. On September 9, 2011, the Commission issued an Order to the ASLB stating that it was evenly split on whether to uphold or overturn the ASLB's decision. The Order also directed the ASLB to "Complete all adjudicatory activities by the end of the fiscal year [September 30, 2011]."

Going forward, the agency will continue to implement oversight programs in regard to radiological waste at sites across the Nation.

Issue: Maintaining flexibility to address regulatory challenges related to the management of spent nuclear fuel and high-level waste.

Action: NRC developed a *Plan for Integrating Spent Nuclear Fuel Regulatory Activities* to address future regulatory challenges related to the management of spent nuclear fuel and high-level waste. This plan is intended to assist NRC in addressing ongoing revisions to the national strategy for ensuring public health and safety and the environment in managing spent nuclear fuel and high-level waste. By coordinating the approach for regulation of spent nuclear fuel and high-level waste storage, potential reprocessing, transportation, and disposal, NRC can improve the efficiency and effectiveness of NRC regulatory processes and provide stability and predictability for stakeholders in a dynamic environment.

Yucca Mountain Crest –
North View. Source:
Bing Images

Issue: Address increasing quantities of high-level radiological waste requiring interim storage or permanent disposal.

Action: To provide technically based safety reviews of license amendment requests to allow credit for the reduction of reactivity due to reactor exposure in spent nuclear fuel, the NRC is conducting a study on the application of burnup credit in light water reactors. The study will address storage in both spent fuel pools and casks, as well as transportation considerations. The results of the study should provide NRC with an independent methodology for the application of burnup credit in spent fuel criticality applications.

Commercial Spent Fuel Pool.
Source: NRC NUREG-1925, Rev 1

NRC also developed and implemented a decisionmaking tool for adopting a graded approach in terms of the depth and rigor of review of applications for certificates of compliance for dry cask waste storage systems. NRC used staff experience and risk insights resulting from a pilot probabilistic risk analysis for dry cask storage systems to build the decisionmaking tool.

Issue: Oversight of low-level waste storage and disposal, including low-level radioactive waste disposal sites. All current low-level waste disposal sites are regulated by Agreement States.

Action: NRC is working to improve the regulatory framework for low-level waste, used by both NRC and Agreements States.

Following are examples of NRC activities:

- NRC completed the Regulatory Basis and most of the proposed rulemaking for requiring a site-specific performance assessment for low level waste disposal sites to demonstrate compliance with the performance objectives in the Code of Federal Regulations.

- NRC also held public meetings on issues related to low-level waste.

- NRC issued a summary of existing guidance to Agreement States for reviewing large-scale low-level radioactive waste blending proposals. The summary of current NRC policy and guidance should assist the Agreement States in making informed decisions until formal rulemaking and guidance is completed.

Low-Level Waste Containers. Source: Bing Images

Issue: Oversight of nuclear waste issues associated with the decommissioning and cleanup of nuclear reactor sites and other facilities.

Action: NRC oversees 10 power reactors, 2 early demonstration reactors, and 12 research and test reactors in safe storage or currently undergoing decommissioning. NRC's regional offices conduct periodic inspections to include reviews of decommissioning nuclear waste management.

Inspector General's Assessment of he Most Serious Management and Performance Challenges Facing NRC

CHALLENGE 5
Implementation of information technology and information security measures.

NRC needs to continue upgrading and modernizing its information technology and security capabilities both for employees and for public access to the regulatory process. Recognizing the need to modernize, the Office of Information Services established goals to improve the productivity, efficiency, and effectiveness of agency programs and operations, and enhance the use of information for all users inside and outside the agency. NRC also needs to ensure that system security controls are in place to protect the agency's information systems against misuse.

The issues related to this challenge and the agency's actions to address each issue include the following:

Issue: Upgrade and manage information technology activities to improve the productivity, efficiency, and effectiveness of agency programs and operations.

Action: NRC continued to work with its business system owners on analysis of legacy applications for future technology modernization, including development of a funding strategy.

Action: The agency awarded a new Information Technology Infrastructure and Support Services contract that will replace the current contract and vendor. The agency expects this will provide a more flexible vehicle for providing information technology services.

Action: An agencywide Virtual Meeting capability was put into place contractually for Fiscal Year 2012.

Action: All users received the full version of Adobe Acrobat Professional. This added cost and management efficiencies by consolidating support and licensing of the application.

Action: NRC coordinated its planning for the new Three White Flint North (3WFN) headquarters building to ensure efficient, cost-effective, and future-thinking information technology infrastructure requirements are understood and included.

Issue: Provide laptop computers with enhanced functionality, security, and support.

Action: NRC completed the implementation of the base laptop program, which provides several enabling technologies for mobile users and supports the capability to "work from anywhere." This included the implementation and/or enhancement of the following services: mobile desktops, loaner laptops, loaner broadband cards, and Blackberry tethering.

Issue: Ensure that information systems and assets are protected.

Action: NRC has begun to identify and explore additional options (e.g., mobile devices, smart phones, and tablets) to provide agency staff with secure, remote access to agency resources.

Action: NRC has continued to make improvements to the Security Operations Center adding tools for centralized logging and continuous monitoring.

Action: Although NRC established an agencywide cyber security situational awareness capability to enhance visibility of and response to emerging threats to NRC information resources, the agency has yet to fully deploy purchased cyber tools that could strengthen its ability to identify, mitigate, and reduce threats against its information systems infrastructure. Effective implementation and coordination among cognizant NRC offices with internal and external response responsibilities to potential network intrusion attacks provide NRC with enhanced capabilities to respond to such threats.

Action: The agency has issued Homeland Security Presidential Directive-12 identification cards to NRC staff and contractors, and Homeland Security Presidential Directive-12 compliant security equipment has been installed in all NRC facilities.

Issue: Ensure that plans for a cyber security inspection program are developed and implemented.

Action: The staff plans to develop an inspection procedure for conducting cyber security inspections at nuclear power plants and hold training for NRC cyber security inspectors. The inspections are planned to be conducted between calendar years 2012 and 2016.

CHALLENGE 6

Administration of all aspects of financial management and procurement.

NRC management is responsible for meeting the objectives of several statutes, including the Federal Managers' Financial Integrity Act. This act mandates that NRC establish controls that reasonably ensure that (1) obligations and costs comply with applicable law; (2) assets are safeguarded against waste, loss, unauthorized use, or misappropriation; and (3) revenues and expenditures are properly recorded and accounted for. This act encompasses program operational, and administrative areas, as well as accounting and financial management.

NRC's procurement of goods and services must be made with an aim to achieve the best value for the agency's dollars in a timely manner. Agency policy provides that NRC's procurement of goods and services supports the agency's mission; be planned, awarded, and administered efficiently and effectively; and be consistent with sound business practices and contracting principles. Agency efforts are currently focused on the goals of achieving (1) a 21st century acquisition program that uses state-of-the-art acquisition methodologies for acquisition planning, execution, management, and closeout; and (2) an acquisition program that fully integrates with agencywide program and financial planning and execution. The issues related to this challenge and the agency's actions to address each issue are as follows:

Financial Management

Issue: Improve the performance and functionality of the agency's new core financial system.

Action: The agency deployed the Financial Accounting and Integrated Management Information System (FAIMIS) on October 1, 2010. FAIMIS replaced five core financial systems with a single Web-based commercial-off-the-shelf software system. While the agency deployed the system as scheduled, there have been a number of performance and functionality issues.

Specifically:

- Delays in issuing license fee bills.
- Frequent system outages prevent staff from entering data into the system in a timely manner.
- Reports used by agency staff to reconcile budget and accounting information are untimely and difficult to interpret.

NRC anticipated initial operational challenges characteristic of new enterprise-wide system deployments. NRC has employed change management and organizational communication strategies to try to reduce the impact of these challenges. For example, the agency has created workgroups to address user concerns, monitored issues through its help desk, and provided additional system training to staff.

Issue: Find a new service provider to host FAIMIS.

Action: The Department of Interior National Business Center, the current host, notified the agency that after September 30, 2012, it would no longer host FAIMIS. The agency is working to find a new service provider before the beginning of Fiscal Year 2013.

Issue: Upgrade the Time and Labor system to a modern, Web-based, user-friendly system.

Action: The agency delayed implementation of plans to upgrade the Time and Labor system in July 2010 because of performance issues identified during production testing. In September 2010, the agency created an Integrated Project Team to perform a technical evaluation and develop a plan for launching a reliable system. The upgrade would provide a modern, Web-enabled version of the existing PeopleSoft Time and Labor software. While this is an upgrade in versions, there are significant differences between the two versions that make the effort similar to a new implementation. The upgraded system will provide increased security, is employee managed, paperless and allows for electronic workflow and electronic signature approval.

Action: The Integrated Project Team took a three-phased approach to implement the new module.

Phase One: Included analyzing and testing application recommendations, and rebuilding the infrastructure in all environments.

Phase Two: Included testing the system for functionality and performance in the rebuilt infrastructure.

Phase Three: Includes user acceptance testing and employee training.

Instructor-led and Web-based training is scheduled to be available for NRC employees in September 2011. The agency developed a SharePoint site to provide a central location for sharing information with NRC employees on progress and for addressing employee questions or issues. The Office of the Chief Financial Officer provided periodic updates on the modernization progress. Present plans call for implementation of the new system on October 23, 2011, in time for processing pay period number 23.

Issue: Respond to the flat or declining budget environment.

Action: As with many other Federal agencies, NRC has had to meet its mission in the face of substantial budget cuts that have flattened the agency's budget. NRC reorganized its budget structure in an effort to improve transparency and target areas of inefficiencies. NRC also established a Transforming Assets into Business Solutions Task Force to analyze and assess NRC business practices and develop cost effective, and cost conscious solutions to manage overhead costs. The task force developed recommendations, which included streamlining the budget formulation process, centralizing human resource management functions, improving contract management, and standardizing information technology solutions.

Inspector General's Assessment of the Most Serious Management and Performance Challenges Facing NRC

Procurement

Issue: Respond to Commission direction to implement a 21st century acquisition program that will consider broader agency programmatic requirements with a more integrated and informed acquisition planning approach that leverages agency resources, programmatic requirements, and contract dollars.

> **Action:** In March 2011, the agency established the Associate Directorate for Strategic Acquisition (ADSA) within the Office of Administration to develop and implement a strategic acquisition program for the agency to address identified process, automation, and workforce skill issues.

> **Action:** ADSA is responsible for procuring and implementing an acquisition module that is compatible with FAIMIS. The acquisition module will standardize and centralize acquisition processes agencywide and automate lifecycle procurement activities for commercial contracts, DOE laboratory agreements, interagency agreements, financial assistance awards, and purchase card transactions within a single business application solution. The agency plans to implement the acquisition module in two phases. However, due to budget constraints and Commission concern over NRC's recent experience with deployment of the FAIMIS Core Financial system, the agency has been unable to finalize an implementation schedule for the acquisition module.

> **Action:** ADSA staff completed a plan of action designed to economize, streamline, and standardize the NRC procurement process through the application of enterprise spend management principles.

> **Action:** NRC recently completed its first comprehensive spend-analysis and identified 10 major portfolio categories as candidates for savings through enterprise sourcing efforts. An ongoing OIG evaluation is assessing NRC's contract award process.

Issue: Implement improvements in the agency's procedures for awarding, negotiating, and managing agreements with DOE laboratories.

Inspector General's Assessment of the Most Serious Management and Performance Challenges Facing NRC

Action: In response to an OIG audit,[4] NRC is in the process of updating Management Directive 11.7, *NRC Procedures for Placement and Monitoring of Work with the U.S. Department of Energy.* Staff has also issued interim guidance for enhanced oversight of the process for placing work with DOE. The interim guidance:

1. Addresses the need for offices to consider the use of commercial sources through market research as part of the decisionmaking process in choosing to use a laboratory and fully document the results/conclusions in DOE laboratory agreement files.

2. Provides clarification for offices regarding the requirement to fully document the rationale and basis for using a DOE laboratory.

3. Addresses the requirement for when offices shall submit Source Selection Justifications to the Division of Contracts for independent review to ensure that commercial sources are fully considered.

4. Identifies milestones for more robust market research requirements prior to awarding DOE laboratory agreements.

Issuance of this interim guidance provides an opportunity for offices to work with the Office of Administration to collaborate on updating and incorporating revisions to Management Directive 11.7.

---

[4] OIG-10-A-12, *Audit of NRC's Management of Agreements with Department of Energy Laboratories* (April 23, 2010)

Issue: Manage the agency's expanded grant program to ensure that grants are awarded in a timely manner and NRC personnel who award and administer grants are provided appropriate training.

Action: In response to recommendations from an OIG audit report,[5] the staff established requirements for the content and organization of NRC grant files. Staff also developed a grants SharePoint site, developed a grants training plan, awarded a contract for off-the-shelf grants training for the Division of Contracts and program office staff, and implemented an internal quality control process to assure regulatory compliance.

Action: The staff updated Management Directive 11.6, *Financial Assistance Program*, which was approved by the Executive Director for Operations on September 16, 2011. The staff is also preparing a *Grants Specialist Desk Guide* for NRC Division of Contracts and program office staff. The purpose of the desk guide is to ensure consistency in the grants process.

---

[5] "OIG-09-A-16, *Audit of NRC's Grant Management Program* (September 29, 2009)

CHALLENGE 7

Managing human capital.

For several years, NRC experienced significant growth resulting from an increased interest in nuclear power. As of July 25, 2011, NRC's workforce is 3,961 staff and it is unlikely that NRC will see any growth over the next several years. Going forward, NRC will need to support increasing mandates within a zero-growth or declining budget environment. NRC must institutionalize an approach that focuses on its mission of protecting the public health and safety while remaining mindful of staff needs. To manage human capital effectively, while continuing to accomplish the agency's mission, NRC must continue to implement initiatives in the following areas:

- Reducing inefficiencies and overhead by centralizing and streamlining processes while maintaining or improving the level of customer service.

- Space planning.

The issues related to this challenge and the agency's actions to address each issue include the following:

Issue: NRC must respond to the flat or declining budget environment.

Action: As NRC transitions from a period of aggressive growth to a nearly flat budget, it is critical that the agency has the most effective organizational structure possible. Despite the current budget pressures on the agency's staffing levels, NRC must continue to look toward the future. Salaries and benefits are a significant driver of the budget and influence how much resources the agency has available for such items as fixed costs, contract support, and travel. To help the agency reshape the workforce, NRC plans to ask the Office of Personnel Management and the Office of Management and Budget for authority to offer some eligible employees in specific situations an opportunity to request voluntary "buy outs" and early retirements. Some of these options are being offered because of changes in technology and the way the agency does business. If approved, these options would be offered some time in the first quarter of Fiscal Year 2012. In addition, the agency has refocused its efforts to manage salaries and benefits costs as well as full-time equivalents.

Issue: NRC must adapt training and development programs to the changing needs of agency staff.

Action: NRC is focusing on a competency-based approach to training to ensure a line-of-sight alignment between employees' learning experiences and the agency's mission. The agency plans to explore and exploit training technologies such as online and distance learning to deliver quality learning opportunities at best cost, when and where they are needed.

Issue: NRC must address knowledge management in light of the high number of senior experts and managers who are or will be eligible to retire.

Action: The agency has in place a variety of human capital strategies to maintain and bolster knowledge and skills during a

period when a large number of experienced staff members are becoming eligible to retire and current and new NRC employees need the benefit of their knowledge. The agency continues to expand and enhance its Knowledge Management program by actively capturing lessons learned from subject matter experts, improving access to lessons-learned and training programs, and building an agencywide Knowledge Center.

Issue: NRC needs to facilitate continuation of its space planning efforts. The first phase of construction of 3WFN, a 14-story building with a 4-level underground parking garage, continues on schedule. Completion of the first phase along with the second phase build out of the interior office space and interior design, is expected by August 2012. When completed, 3WFN will provide office space for approximately 1,400 NRC staff members and allow the agency to reconsolidate headquarters staff who are now dispersed among four offsite locations. There is no funding in the budget for either above ground or underground pedestrian access between One White Flint North and 3WFN. To access 3WFN, agency employees will have to cross Marinelli Road, which is a multi-lane road. NRC faces two challenges related to 3WFN. The agency must ensure that:

- Building requirements are met and within budget.

- Provisions are put in place to ensure safe pedestrian movement between the buildings.

    Action: NRC will (1) review and approve the Construction Drawings for the 3WFN building and the construction bid pricing; (2) complete the interior fit-out of the building; (3) develop

    specifications, procure and install workstations, and (4) relocate approximately 1,400 staff and contractors as well as the Professional Development Center, Data Center, and Headquarters Operations Center to the building. The target date to complete relocation into the 3WFN building remains December 2012. The goal is to provide a quality working environment for NRC employees within the established budget.

    Action: The Montgomery County Department of Transportation and NRC are continuing to work together to maximize pedestrian safety around the White Flint complex. As part of the effort, the Montgomery County Department of Transportation removed a temporary crosswalk on Marinelli Road east of the median wall and constructed a new crossing area near the entrance to White Flint Complex. NRC installed sidewalk barriers along the south side of Marinelli Road to direct pedestrians to the new crossing and to discourage jay walking between the temporary One White Flint North entrance on the south side of Marinelli Road and the Metrorail parking garage on the north side while lobby construction was underway.

Inspector General's Assessment of the Most Serious Management and Performance Challenges Facing NRC

IV.   CONCLUSION

The seven challenges contained in this report are distinct, yet are interdependent to accomplishing NRC's mission. For example, the challenge of managing human capital affects all other management and performance challenges.

The agency's continued progress in taking actions to address the challenges presented should facilitate achieving the agency's mission and goals.

Appendix

SCOPE AND METHODOLOGY

SCOPE

This evaluation focused on the IG's annual assessment of the most serious management and performance challenges facing the NRC. The challenges represent critical areas or difficult tasks that warrant high level management attention. To accomplish this work, the OIG focused on determining (1) current challenges, (2) the agency's efforts to address the challenges during Fiscal Year 2011, and (3) future agency efforts to address the challenges.

METHODOLOGY

OIG reviewed and analyzed pertinent laws and authoritative guidance, agency documents, and OIG reports, and sought input from NRC officials concerning agency accomplishments relative to the challenge areas and suggestions they had for updating the challenges. Specifically, because challenges affect mission critical areas or programs that have the potential to impact agency operations or strategic goals, NRC Commission members, offices that report to the Commission, the Executive Director for Operations, and the Chief Financial Officer were afforded the opportunity to share any information and insights on this subject.

OIG staff conducted this evaluation from May through August 2011 at NRC headquarters.

MANAGEMENT DECISIONS AND FINAL ACTIONS ON OIG AUDIT RECOMMENDATIONS

MANAGEMENT DECISIONS AND FINAL ACTIONS ON OIG AUDIT RECOMMENDATIONS

The agency has established and continues to maintain an excellent record in resolving and implementing audit recommendations presented in OIG reports. Section 5(b) of the Inspector General Act of 1978, as amended, requires agencies to report on final actions taken on OIG audit recommendations. The following table gives the dollar value of disallowed costs determined through contract audits conducted by the Defense Contract Audit Agency and NRC's Office of the Inspector General. Because of the sensitivity of contractual negotiations, details of these contract audits are not furnished as part of this report. As of September 30, 2011, there were no outstanding audits recommending that funds be put to better use.

MANAGEMENT REPORT ON OFFICE OF THE INSPECTOR GENERAL AUDITS WITH DISALLOWED COSTS

For the period October 1, 2010 – September 30, 2011

Category	Number of Audit Reports	Questioned Costs	Unsupported Costs
1. Audit reports with management decisions on which final action had not been taken at the beginning of this reporting period.	0	$0	$0
2. Audit reports on which management decisions were made during this period.	0	$0	$0
3. Audit reports on which final action was taken during this report period.			
(i) Disallowed costs that were recovered by management through collection, offset, property in lieu of cash, or otherwise.	0	$0	$0
(ii) Disallowed costs that were written off by management.	0	$0	$0
4. Reports for which no final action had been taken by the end of the reporting period.	0	$0	$0

MANAGEMENT DECISIONS NOT IMPLEMENTED WITHIN 1 YEAR

For the OIG audit reports listed in the following tables, the NRC made management decisions before October 1, 2010. As of September 30, 2011, NRC had not taken final action, including OIG final review and closure, on some issues. Completion of the activities listed in the column "Actions Pending" will complete agency action on the listed OIG audit and evaluation recommendations. Recommendations with "Actions Pending" from three OIG reports contain security related information that is not released to the public.

GOVERNMENT PERFORMANCE AND RESULTS ACT: REVIEW OF THE FISCAL YEAR 1999 PERFORMANCE REPORT (OIG-01-A-03)

February 23, 2001

The OIG conducted this audit at the request of the chairman of the Senate Committee on Governmental Affairs to determine if NRC's FY 1999 performance data was valid and reliable and if the FY 2000 performance data would be more valid and reliable. The audit found that while NRC was improving and strengthening its performance reporting process, as interim policy guidance, the agency needed to institute management control procedures to produce valid and reliable data. The agency should then institutionalize the procedures in an NRC Management Directive (MD).

Open Recommendations	Actions Pending
1. Develop an NRC management directive (MD) to provide the management controls needed to ensure that the NRC produces credible Government Performance and Results Act (GPRA) documents.	The NRC is in the process of revising Management Directive 4.4, Internal Control, and expects to issue the final version by the end of Fiscal Year 2012. The NRC also issues agency guidance and instructions annually for completing GPRA documents, including establishing performance metrics and reporting on unmet goals.
3. Include guidance on reporting unmet goals in both the management directive and the interim policy guidance on implementing GPRA initiatives.	This recommendation will be addressed by replacing Management Directive (MD) 4.7, NRC Long Range Planning, Programming and Budget Formulation, and Handbook with three separate MDs, "Strategic Planning Process, (new), Budget Formulation, (revision to MD 4.7, and Performance Management (new). The MD for Performance Management, including guidance for reporting on unmet metrics will be developed following the completion of a formal "Business Process Improvement" (BPI) project, which is expected to eliminate unnecessary metrics, perform a "vertical slice" approach that involves both program and corporate metrics, and then conduct a broader study culminating an integrated performance management system. The BPI project is scheduled to be completed in the 3rd quarter of FY 2012, with issuance of a draft MD on Performance Management in the 4th quarter of FY 2012.

AUDIT OF NRC'S REGULATORY OVERSIGHT OF SPECIAL NUCLEAR MATERIALS (OIG-03-A-15)

May 23, 2003

The OIG conducted this audit to determine whether NRC adequately ensures that its licensees control and account for special nuclear material (SNM). The audit found that NRC's current levels of oversight of licensees' material control and accounting (MC&A) activities do not provide adequate assurance that all licensees properly control and account for SNM. The audit reported that NRC performs limited inspections of licensees' MC&A activities and cannot assure the reliability of data in the Nuclear Materials Management and Safeguards System. The U.S. Department of Energy manages this computer database and shares it with the NRC as the national system for tracking certain private- and Government-owned nuclear materials.

Open Recommendations	Actions Pending
1. Conduct periodic inspections to verify that material licensees comply with MC&A requirements, including but not limited to visual inspections of licensees' SNM inventories and validation of report information. 3. Document the basis of the approach used to risk-inform NRC's oversight of MC&A activities for all types of materials licensees.	In SECY-05-0143, the staff recommended that the Commission approve the staff's proposed enhancements to the Material Control and Accounting (MC&A) regulations, inspection program, and licensing process. Consistent with information provided in previous status reports, in response to the associated staff requirements memorandum (SRM) to SECY-05-0143 dated November 18, 2005, the staff completed the development of the technical basis for the 10 CFR Part 74 rulemaking. This technical basis addressed the requirement to risk inform the MC&A requirements and the need to complete documentation of the basis for risk informing the MC&A program. This was described in the MC&A rulemaking plan (SECY-08-0059) dated April 25, 2008. The SRM SECY-08-0059 was issued on February 5, 2009. The Commission approved the staff's rulemaking Option 4, directing the staff to revise and consolidate current MC&A regulations into 10 CFR Part 74, "Material Control and Accounting of Special Nuclear Material." The proposed rule package to amend 10 CFR Part 74 was submitted to the Commission on September 14, 2011. This revision of 10 CFR Part 74 will complete the OIG recommendations 1 and 3.

AUDIT OF THE BUDGET FORMULATION PROCESS (OIG-05-A-09)

January 31, 2005

OIG conducted the audit to determine whether the budget formulation portion of the NRC's planning, budgeting, and performance management process is effectively used to develop and collect data to align resources with strategic goals and is efficiently and effectively coordinated with program and support offices.

Open Recommendations	Action Pending
1. Clarify the roles and responsibilities of the Chief Financial Officer and the Executive Director for Operations in the budget formulation process.	The recommendations are being addressed as part of the revision to Management Directive (MD) and Handbook 4.7, *NRC Long Range Planning, Programming and Budget Formulation*. The agency has modified its approach to the replacement of the MD and Handbook in order to improve policy communication, organization and achieve agency consensus on the policies covered. The document will be replaced with three separate MDs: *Strategic Planning Process* (new); *Budget Formulation* (MD 4.7); and, *Performance Management* (new). The three recommendations will be addressed in the revised MD 4.7, *Budget Formulation*.
2. Document the decisionmaking process and the roles and responsibilities of the program review committee.	
3. Document the budget formulation process to ensure a logical, comprehensive sequencing of events that provides for obtaining early Commission direction and approval.	

Audit of NRC's Telecommunications Program (OIG-05-A-13)

June 7, 2005

OIG conducted this audit to evaluate controls over the use of NRC telecommunications services and the physical security of NRC telecommunications systems. OIG found that the agency needs to strengthen controls over the use of telecommunications services and the physical security of NRC telecommunications systems.

Open Recommendation	Action Pending
3. Revise Management Directive 2.3 (MD 2.3) and Handbook, "Telecommunications," to include effective management controls over NRC headquarters staff use of agency telecommunications services.	In September 2011, the staff provided the revised version of MD 2.3 and Handbook, including the recommended change, to the Executive Director for Operations for final review and signature.

AUDIT OF NRC'S DECOMMISSIONING PROGRAM (OIG-05-A-17)

September 21, 2005

OIG conducted this audit to determine whether the NRC's decommissioning program achieves desired performance results, as stated in the Strategic Plan and reported in the Performance and Accountability Report.

Open Recommendation	Action Pending
1. Clarify and disseminate expectations for generating and maintaining supporting documentation for performance data to staff responsible for preparing and collecting performance data.	The recommendation is being addressed as part of the revision to Management Directive (MD) and Handbook 4.7, *NRC Long Range Planning, Programming and Budget Formulation*. The agency has modified its approach to the replacement of the MD and Handbook in order to improve policy communication, organization and achieve agency consensus on the policies covered. The document will be replaced with three separate MDs: *Strategic Planning Process (new); Budget Formulation* (MD 4.7); and, *Performance Management* (new). The recommendation will be addressed in the revised MD 4.7, *Budget Formulation*.

AUDIT OF NRCs REGULATION OF NUCLEAR FUEL CYCLE FACILITIES (OIG-07-A-06)

January 10, 2007

This audit determined whether the NRC has an effective and efficient approach to fuel cycle facility oversight. The audit found that the NRC could enhance the current Fuel Cycle Facility Oversight Program by developing and implementing a framework modeled after a structured process, such as the Reactor Oversight Process (ROP).

Open Recommendation	Action Pending
1. Fully develop and implement a framework for the Fuel Cycle Facility Oversight Program (FCFOP) that is consistent with a structured process, such as the Reactor Oversight Process (ROP).	The staff continues to work on enhancements to the fuel cycle oversight process (FCOP). This effort has included extensive interactions with stakeholders to solicit their input on enhancements to the FCOP. In accordance with SRM-10-0031, the staff will provide the Commission a paper that will recommend an FCOP framework that provides a structure that is similar to the ROP. In addition, the staff will provide two alternatives for Commission consideration. The alternatives provide incremental enhancements to the existing FCOP. After a Commission briefing on this subject, the staff expects to receive direction on how to proceed.

AUDIT OF NRC'S PLANNED CYBERSECURITY PROGRAM (OIG-08-A-06)

March 18, 2008

This audit determined how upcoming changes to the NRC's cybersecurity oversight processes might impact the agency's physical security inspection program

Open Recommendation	Actions Pending
1. Develop and implement plans for a cybersecurity oversight program that captures skill set and workload requirements for cybersecurity inspections, and targets resources to prepare for program implementation in calendar year 2010.	Through the planning, budgeting, and performance management process, one Cyber Security Specialist FTE was hired at the end of FY 2010. Two additional FTEs will be assigned to inspection oversight activities at the beginning of FY 2012 to continue program development in FY 2012 and FY 2013. With these resources, the staff will augment the continued development of a temporary instruction, related enforcement guidance, a training and qualification program, and workshops and pilots planned for CY 2012. The staff developed and conducted a pilot cyber security training course for the Cyber Security Inspectors in January 2011. The staff anticipates conducting a second cyber-security training course in late CY 2012 to prepare additional cyber security inspectors and support staff for upcoming inspections.

The inspections are planned to be conducted between CY 2012 and CY 2016 using the temporary instruction which will provide the framework for further development of the cyber security oversight program, and the program's transition into the agency's reactor oversight process. |

AUDIT OF NRC'S ACCOUNTING AND CONTROL OVER TIME AND LABOR REPORTING OIG-08-A-11

June 17, 2008

The objectives of the audit were to determine whether the NRC established and implemented internal controls over time and labor reporting to provide reasonable assurance that hours worked in pay status and hours absent are properly reported and that the time and labor system is easy and efficient to use.

Open Recommendations	Actions Pending
3. The CFO should conduct a detailed system analysis and eliminate redundant paper forms that are not needed.	In Pay Period 23 – 2011 OCFO deployed the HRMS Modernization system. This system eliminated paper forms using electronic workflow for all approvals. The Summary Approval Report (SAR), all leave request forms, time & labor unit transfer forms and security request forms are all part of the electronic workflow.
4. The CFO should use electronic signatures for time reporting and approval.	With the deployment of the HRMS Modernization system in Pay Period 23 – 2011, one of the enhancements was electronic workflow. Electronic workflow would allow for electronic signatures for time and labor approval.

AUDIT OF NRC'S PREMIUM CLASS TRAVEL (OIG-08-A-16)

September 12, 2008

The objectives of the audit were to determine whether travel costs associated with premium air travel (i.e., per diem) are properly authorized, justified, and documented and to determine whether premium air travel is properly authorized, justified, and documented. OIG specifically assessed compliance with requirements in OMB Memorandum M-08-07.

Open Recommendation	Action Pending
1. Update Management Directive 14.1 to clearly identify premium travel authorizing officials; clarify "Delegation of Authority: and require this to be in written form; and clarify the 14 hour rule, specifically the rest period.	Management Directive 14.1 "Official Temporary Duty Travel" has been revised to incorporate these changes and sent to the Offices for comments. Comments have been received and are being evaluated by the Office of the Chief Financial Officer (OCFO). We expect to have Management Directive 14.1 "Official Temporary Duty Travel" updated and reissued with a target completion date of January 31, 2012.

MANAGEMENT DECISIONS AND FINAL ACTIONS ON OIG AUDIT RECOMMENDATIONS

AUDIT OF NRC'S ENFORCEMENT PROGRAM (OIG-08-A-17)

September 26, 2008

The objective of the audit was to review the NRC's enforcement program to determine whether the program is comprehensive and consistently implemented and whether enforcement decisions are based on complete and reliable data. OIG identified that the regional offices implement the enforcement program inconsistently because the agency has not issued clear and comprehensive guidance to facilitate the program. In addition, the audit identified that information used for decision making and reporting purposes is not complete and reliable.

Open Recommendations	Actions Pending
2. Define systematic data collection requirements for non-escalated enforcement actions.	The NRC staff is currently developing a Web-based licensing (WDL) system that will track non-escalated enforcement actions issued to materials licensees. The database is expected to be available for enforcement data collection in mid-2012. The staff has evaluated the capabilities available with the reactor program system (RPS) and determined that it is a sufficient tool for tracking and trending non-escalated reactor enforcement actions.
3. Develop and implement a quality assurance process that ensures that collected enforcement data is accurate and complete.	Actions to address Recommendation 3, which involve the development of procedures for data entry and auditing of WBL, will follow the actions to address Recommendation 2. In the interim, the NRC staff completed a follow-up audit of NRC Form 591s that resulted in a recommendation for action assuring alignment to one version of the form.

INDEPENDENT EVALUATION OF NRC'S IMPLEMENTATION OF THE FEDERAL INFORMATION SECURITY MANAGEMENT ACT FOR FISCAL YEAR 2008 (OIG-08-A-18)

September 26, 2008

The objective of this review was to perform an independent evaluation of the Nuclear Regulatory Commission's implementation of the Federal Information Security Management Act (FISMA) for FY 2008.

Open Recommendation	Action Pending
4. Develop a process for verifying that all Federal Desktop Core Configuration (FDCC) controls are implemented for all desktop and laptop computers, including both those that are centrally managed under the agency's seat management contract and those that are owned by the agency regardless of whether or not they are connected to the agency's network.	The staff will use Secure Content Automation Protocol (SCP) and the FDCC auditing tools to verify that the agency is compliant with M-08-22 for both Office of Information Services centrally managed and Region/Program Office managed computer assets. The staff is scheduling an FDCC compliance check on all networked endpoints using an installed NIST certified network appliance and sample networked and non networked endpoints using NIST certified off network stand alone FDCC approved tool in the first and second quarter of FY 2012.

AUDIT OF THE COMMITTEE TO REVIEW GENERIC REQUIREMENTS (OIG-09-A-06)

February 2, 2009

The Office of the Inspector General (OIG) of the U.S. Nuclear Regulatory Commission (NRC) conducted this audit to determine if the Committee to Review Generic Requirements (CRGR) adds value for the Executive Director for Operations' decisionmaking purposes and whether the committee's function is still needed.

Open Recommendation	Actions Pending
1. Develop, document, implement, and communicate an agency-wide process for reviewing backfit issues to ensure that generic backfits are appropriately justified based on NRC regulations and policy.	In addressing Recommendation 1 and in its role of providing CRGR support, the staff coordinated the implementation of an Action Plan with the relevant offices and regions. The planned activities are currently envisioned to include at least the following five areas: (1) revise the CRGR Charter, (2) revise Management Directive (MD) 8.4, "Management of Facility-specific Backfitting and Information Collection", (3) develop office and regional procedures that are consistent with the revised MD 8.4, (4) develop an agency-wide Web-based backfit training program, and (5) document, communicate, and implement an overarching agency-wide backfit program. The CRGR and Office staff worked together to establish a centralized agency resource for backfit training.
	On March 8, 2011 the CRGR was signed by the OEDO and issued. The CRGR staff is in the final stages of addressing program office comments and plans on providing MD 8.4 in October, 2011, for finalizing and preparation for issuance in November, 2011. Once the MD 8.4 is issued, the offices and regions will ensure their respective procedures are consistent with the revised MD. These revisions will address important elements for ensuring effective overarching management of generic and plant-specific backfits.
	At the present, CRGR staff, the Office of the General Counsel and the Office of Human Resources are engaged in the process of reviewing and updating a previous draft of an agencywide Web-based backfit training program. An additional step will be to develop a training module on the overall process and then to develop program-specific modules that can be used by the program offices and regions, as appropriate.
	These planned activities will document the role of the CRGR and the staff process for ensuring compliance with backfit requirements and procedures that have evolved since the inception of the CRGR. The CRGR plans to communicate the changes to the staff and verify that the relevant offices and regions have incorporated processes to ensure backfit rules and requirements are followed.
	The projected completion date for this recommendation is December 31, 2011.

AUDIT OF NRC'S AGREEMENT STATE PROGRAM (OIG-09-A-08)

March 16, 2009

The audit objective was to assess NRC's oversight of the adequacy and effectiveness of Agreement State programs. OIG focused its review on the Integrated Materials Performance Evaluation Program (IMPEP) process as well as other elements of the Agreement State program. OIG identified program adequacy and effectiveness issues that require management's attention.

Open Recommendation	Action Pending
4. Develop a standardized data collection process that can be used as the basis of an information sharing tool on a national level.	Based on input from the Agreement States, NRC staff concluded that a voluntary, password protected database on the FSME public Website that includes data involving bans, revocations, suspensions, and denials is the most suitable option. NRC staff is working towards the development of a national enforcement database to host the information. Agreement States have voluntarily begun contributing information.

AUDIT OF THE NRC'S WAREHOUSE OPERATIONS (OIG-09-A-09)

March 31, 2010

The purpose of this audit was to determine whether NRC has established and implemented an effective system of internal controls for maintaining accountability and control of agency property stored in the warehouses.

Open Recommendation	Action Pending
2. Conduct the required security survey of the NRC Annex.	The Federal Protective Service (FPS) Area Commander notified the Office of Administration (ADM), Division of Facilities and Security, that a Building Security Assessment of the NRC Annex was completed on August 6, 2010. In actuality, FPS surveyed another NRC warehouse location rather than the Annex. As a result of setbacks due to scheduling delays and new inspection requirements, FPS is not expected to conduct another survey of the NRC Annex until later in fiscal year 2012. A copy of the FPS survey, when completed, will be provided to the Office of the Inspector General. Targeted Completion Date: September 30, 2012.

AUDIT OF NRC'S GRANT MANAGEMENT PROGRAM (OIG-09-A-16)

September 29, 2009

The audit objective was to determine whether NRC has established and implemented an effective system of internal controls for grants management.

Open Recommendations	Actions Pending
1. Resolve outstanding Lean Six Sigma issues, including definition of the competitive grant process, roles and responsibilities, development of a shared electronic grant database, and scope of Office of Small Business and Civil Rights reviews.	The Office of Administration (ADM) resolved this recommendation in part by incorporating Lean Six Sigma recommendations in Management Directive (MD) 11.6, "Financial Assistance Program." The MD was revised to include the definition of the competitive grant process, roles and responsibilities, and the scope of the Office of Small Business and Civil Rights reviews. On September 16, 2011, this MD was published. With respect to the development of a shared electronic database, ADM developed a SharePoint site for grants management which includes an improved document/reference library. This recommendation is completed pending review and closure by OIG.
2. Update Management Directive 11.6 to comprehensively address NRC's competitive and non-competitive grant program, including (a) roles and responsibilities of individuals and offices involved in the grant process, (b) process for awarding grants, and (c) required monitoring by project officers.	ADM revised MD 11.6 through the formal MD process to comprehensively address NRC's competitive and non-competitive grants program, including (a) roles and responsibilities of individuals and offices involved in the grants process, (b) process for awarding grants, and (c) required monitoring by Project Officers. On September 16, 2011, this MD was published. This recommendation is completed pending review and closure by OIG.
5. Ensure that staff working on grants complete the required training within specified timeframe identified in response to recommendation 4.	Grant staff are expected to complete the training identified in NRC's Grant Management Certification Training Program by December 31, 2011, consistent with the response to recommendation 4, which states "Develop grant specific training requirements for staff who work on grants to include a reasonable period of time (such as 18 months) for completion of the training." The training is being monitored by the NRC Acquisition Career Manager in coordination with the Office of Human Resources. Targeted completion date: December 31, 2011.

AUDIT OF NRC'S MANAGEMENT OF AGREEMENTS WITH DEPARTMENT OF ENERGY LABORATORIES (OIG-10-A-12)

April 23, 2010

The audit objective was to determine whether NRC has established and implemented an effective system of internal control over the placement and monitoring of work with DOE Laboratories.

Open Recommendations	Actions Pending
1. Revise MD 11.7 to require NRC offices to consider the use of commercial sources through market research (e.g., sources-sought announcements, procurement history, expert knowledge) as part of the decision-making process in choosing to use a lab, and fully document the results/conclusions in the agreement files.	ADM is in the process of revising Management Directive (MD) 11.7. The updated version of the MD includes language that requires offices to consider the use of commercial sources through market research as part of the decision making process in choosing to use a DOE lab. The revised MD also provides guidance for fully documenting the results/conclusions of the market research in the agreement files. The revised MD 11.7 will be published by January 31, 2012.
2. Clarify MD 11.7 to emphasize the requirement to fully document the rationale and basis for using a DOE lab.	The revision to MD 11.7 will emphasize the requirement to fully document the rationale and basis for using a DOE lab. The revised MD 11.7 will be published by January 31, 2012.
3. Revise MD 11.7 to require independent review of justifications by DC [Division of Contracts] personnel to ensure that commercial sources were fully considered.	ADM has prepared review criteria, which incorporates input from the program offices, that will be used in determining which justifications must be reviewed by DC prior to award of an agreement to DOE. Those justifications that do not meet the criteria will continue to be reviewed by the Associate Competition Advocate in each office. The revised MD 11.7 will include the new review criteria. The revised MD 11.7 will be published by January 31, 2012.
7. Issue a delegation of authority to OIP to award, extend, modify, and terminate DOE lab agreements.	This recommendation is considered to be completed via the Senior Procurement Executive's memo to the Office of International Programs dated April 6, 2011, delegating them the contractual authority to award, extend, modify, and terminate interagency agreements, including Department of Energy Laboratory Agreements. Closure is pending submission of the scheduled periodic status memo on this OIG report's open recommendations and the OIG's subsequent review.

AUDIT OF NRC'S TELEWORK PROGRAM (OIG-10-A-13)

June 9, 2010

The audit objectives were to determine NRC's readiness to have staff telework under emergency situations, the adequacy of internal controls associated with the telework program, and if NRC's telework program complies with relevant law and Office of Personnel Management guidance.

Open Recommendations	Actions Pending
5. Reference emergency planning and information technology procedures in telework guidance. 6. Develop and implement a Management Directive and Handbook for the telework program. 8. Develop and implement a procedure for assessing and reporting the results of full-time telework arrangements to the Office of Human Resources.	A telework Management Directive (MD) and Handbook and a standard operating procedure for telework have been drafted. Emergency planning and information technology will be referenced in the Handbook and in the standard operating procedure for telework. The Handbook and standard operating procedure are scheduled to be issued with the MD in September, 2012. The telework MD and Handbook have been drafted. The telework MD and Handbook are scheduled to be issued in September, 2012. Guidance on assessing and reporting the results of a full time telework agreement will be included in a telework standard operating procedure which is scheduled to be issued with the MD. The telework MD and Handbook are scheduled to be issued in September, 2012.

AUDIT OF NRC'S PROCESS FOR CLOSED MEETINGS (OIG-10-A-14)

June 9, 2010

The objective of the audit was to determine if the NRC's process for conducting meetings that are closed to the public hinders the transparent transaction of nuclear regulation.

Open Recommendations	Action Pending
1. Revise MD 3.5 to enhance NRC's closed staff meeting process. Specifically,	Management Directive (MD) and Handbook 3.5, "Attendance at NRC Staff-Sponsored Meetings" and ADAMS Template NRC-001, were revised to address the recommendations from the OIG report and were provided to staff for review and comment on November 19, 2010. Second drafts of these documents were again provided for comment on April 25, 2011. The MD and Handbook have been provided to the NRC's Office of Information Services as part of the official process to finalize these documents.
a. Clearly define what constitutes a "meeting."	
b. Clarify guidance to ensure that closed staff meeting notices and summaries are appropriately available to the public through ADAMS.	
c. Revise ADAMS Template NRC-001 and MD 3.5 to ensure that the guidance for preparing closed staff meeting notices is consistent.	
2. Establish a timeframe for issuing closed staff meeting notices and summaries.	

AUDIT OF NRC EMPLOYEE USE OF THE FEDERAL CALLING CARD (OIG-10-A-15)

July 30, 2010

OIG conducted this audit to determine whether the NRC has established and implemented an effective system of internal control over the use of Federal calling cards. OIG identified cost-effective measures that NRC could use to (1) improve management controls over the agency's Federal calling card program and (2) provide information to employees about cost-effective calling options. By imposing a basic level of cost-effective oversight over the calling card program, the agency could increase its confidence that staff are using this resource as intended.

Open Recommendations	Actions Pending
1. Develop and implement a plan to assess the validity of high calling card usage on a periodic basis.	As part of the transition of telecommunications services from the FTS2001 contract vehicle to the Networx contract vehicle, the staff is currently replacing/re-issuing calling cards. The Networx vehicle provides the ability to assess usage and monitor inventories at a more detailed level than is currently available and also provides an opportunity to reassess calling card inventory and requirements. The transition to Networx calling cards was completed by September 30, 2011. After the transition was completed, the staff began work to develop and implement a plan to evaluate and identify high calling card usage on a monthly basis. This plan will be implemented by December 31, 2011.
2. Develop and implement a policy to conduct annual inventories of calling cards and reconcile differences in a timely manner.	Management Directive (MD) 2.3 which is currently under revision contains requirements regarding the maintenance of an accurate inventory of telecommunications devices and services. As part of the transition of telecommunications services from the FTS2001 contract vehicle to the Networx contract vehicle, the staff replaced/re-issued calling cards. The Networx vehicle gives the staff the ability to assess usage and monitor inventories at a more detailed level than is currently available and also provides an opportunity to reassess calling card inventory and requirements. The transition to Networx calling cards was completed by September 30, 2011. Following the transition, the staff began development of an internal process to conduct inventories of calling cards and reconcile any differences. The development and implementation of this process is scheduled to be completed by December 31, 2011.

Open Recommendation	Actions Pending
3. Calculate the relative costs of various calling options and, if there are significant differences, communicate information to staff about cost-effective calling options.	Following the transition to Networx calling cards, the staff began reviewing the costs of calling options. If there are significant cost savings identified, the staff will communicate findings to the CIO by December 31, 2011. In addition, after the transition to Networx calling cards, the staff began development of an internal process to conduct inventories of calling cards and reconcile any differences. The development and implementation of this internal process is scheduled to be completed by December 31, 2011.

AUDIT OF NRC'S OVERSIGHT OF IRRADIATOR SECURITY (OIG-10-A-17)

September 2, 2010

The audit objective was to evaluate the adequacy of NRC's oversight of the industrial irradiator security. The original audit scope was expended beyond irradiators to address radioactive materials security program as a whole. OIG report made three recommendations to improve the agency's radioactive materials security program.

Open Recommendations	Actions Pending
1. Reevaluate and determine the frequency of security inspections based on a risk-informed approach.	FSME is in the process of implementing a pilot program to reevaluate security inspection frequency for licensees that possess risk significant radioactive material to determine if more frequent inspection would result in greater compliance with security requirements. The pilot program focuses on self-shielded irradiators, which are now inspected at a Priority 5 frequency. It will provide risk insights that will indicate how a licensee performs with the security requirements at roughly twice their current inspection frequency (i.e. going from every 5 years to every 2 or 3 years).
3. Develop and offer periodic refresher training for all individuals involved in the materials security program.	Refresher training for staff involved in the materials security program will be developed; the periodicity of the refresher training will depend on changes to security requirements. As an example, the staff intends to develop refresher training after the proposed security rulemaking is finalized.
	FSME established a working group, which includes Agreement State participation, to revise and update IMC 1246, "Formal Qualification Programs in the Nuclear Material Safety and Safeguards Program Area." The working group will review and make recommendations related to refresher training based on the overall efficiency and effectiveness of the qualification programs for radioactive materials license reviewers and inspectors.

AUDIT OF THE U.S. NUCLEAR REGULATORY COMMISSION'S VENDOR INSPECTION PROGRAM (OIG-10-A-20)

September 28, 2010

OIG conducted an audit of the implementation of the NRC's vendor inspection program. The objective of the audit was to assess NRC's regulatory approach for ensuring the integrity of domestic and foreign safety-related parts and services supplied to current or prospective nuclear power reactors.

Open Recommendations	Actions Pending
1. Develop an Office of New Reactors (NRO) Vendor Inspection Program planning document that: a. Articulates a clear purpose for the Vendor Inspection Program; and b. Establishes metrics to evaluate the success of the Vendor Inspection Program.	NRO staff is authoring a planning document that will include key program planning elements. The plan will establish an overall strategy, goals, and methodologies for setting priorities, identifying performance metrics, and managing resources. NRO staff will include program-level metrics in this program planning document. NRO staff has held planning meetings and management briefings for the development of such a program planning document. NRO staff plans to incorporate the elements of its actions in response to OIG's recommendations into the program plan. For example, the methodologies for identifying vendors and selecting vendors for inspection and the strategy for outreach and communications will be included in the program plan. The staff will issue program plan by December 30, 2011. The OIG status for this recommendation is "resolved."
2. Develop and document a methodology to identify vendors that supply safety-related parts and services to the nuclear industry with Appendix B quality assurance programs.	The NRC staff is developing a methodology to identify vendors that supply safety-related parts and service to the nuclear industry. The staff will issue methodology by October 28, 2011. The OIG status for this recommendation is "resolved."
3. Develop and document a risk-informed methodology to select vendors for inspection.	The NRC staff is developing a risk-based methodology to select vendors for inspection. The staff will issue methodology by October 28, 2011. The OIG status for this recommendation is "resolved."
4. Develop and use a vendor outreach/communications plan.	The NRC staff developed a strategy for enhanced vendor outreach and communications. The strategy is being implemented. The current revision of the strategy will be incorporated into the program planning document discussed in Recommendation 1. The OIG status for this recommendation is "resolved."

Open Recommendations	Actions Pending
5. Align NRC guidance and regulations to clarify acceptance methods for commercial-grade dedication.	The NRC staff has been actively pursuing efforts to develop a holistic solution to 10 CFR Part 21 and associated issues, including commercial-grade dedication. This has included internal discussions, management briefings, formation of the 10 CFR Part 21 Working Group, and presentations at industry meetings and the 2011 Regulatory Information Conference.
	On September 29, 2011, the NRC staff issued a Commission paper informing the Commission of the staff's plans to develop the regulatory basis to clarify 10 CFR Part 21. The OIG status for this recommendation is "resolved."
6. Issue regulatory guidance to clarify sampling expectations for commercial-grade dedication.	As noted above for Recommendation 5, the NRC staff issued a Commission paper for rulemaking on the topics associated with 10 CFR Part 21, including the sampling expectation for commercial-grade dedication. The OIG status for this recommendation is "resolved."
7. Issue regulatory guidance describing a process that NRC considers acceptable for compliance with Part 21.	As noted above for Recommendation 5, the NRC staff issued a Commission paper for rulemaking on the topics associated with 10 CFR Part 21, including the description of a process it considers acceptable for compliance with 10 CFR Part 21. The NRC staff considers "evaluating and reporting" to be paramount to the clarification of 10 CFR Part 21. This has been the central topic of two of the NRC staff's meetings of the 10 CFR Part 21 Working Group and is a central issue in the Commission paper. The OIG status for this recommendation is "resolved."
9. Develop guidance that clarifies the requirements for vendors on how to approve accredited commercial-grade calibration laboratories for safety-related applications.	The NRC staff plans to author a safety evaluation report (SER) to clarify the requirements for vendors on how to approve accredited commercial-grade calibration laboratories for safety-related applications in response to an industry submittal, expected in 2012. The NRC staff will then issue generic communications to industry indicating that the submittal and SER are available.
	The NRC staff has been extensively involved with industry groups to develop a complete process that can be fully endorsed by the NRC. In addition, the NRC staff included this information in the development of the Commission paper and the subsequent regulatory basis for 10 CFR Part 21 rulemaking and associated guidance. The OIG status for this recommendation is "resolved."

Open Recommendation	Actions Pending
10. Develop and implement a formal agency-wide strategy and plan in order to monitor and evaluate Counterfeit, Fraudulent, and Suspect Items (CFSI).	In late December 2010, the NRC staff began an agency-wide approach to address the threat of CFSI. The NRC staff is working with both internal and external stakeholders to assess the current status and develop an effective path forward. The NRC staff established the following four working groups to engage offices potentially affected by CFSI: 1. Working Group on Supply Chain Oversight 2. Working Group on CFSI Communications 3. Working Group on CFSI Response Protocols 4. Working Group on Cyber Security Supply Chain Oversight This effort is consistent with similar ongoing initiatives within the U.S. Government. The NRC staff is actively participating on Federal working groups to share CFSI-related information, anti-counterfeiting practices, proper Federal response protocols, knowledge, and resources. The NRC staff has begun similar outreach activities with the U.S. Department of Energy, the Electric Power Research Institute, and the Nuclear Procurement Issues Committee. The NRC staff has also made CFSI presentations to the international community and is continuing its outreach activities. The NRC staff is leveraging these diverse activities to develop its formal agency-wide strategy and plan to monitor and evaluate CFSI in a Commission paper. The staff will issue the Commission paper by October 28, 2011. The OIG status for this recommendation is "resolved."

SUMMARY OF FINANCIAL STATEMENT AUDIT AND MANAGEMENT ASSURANCES

SUMMARY OF FINANCIAL STATEMENT AUDIT AND MANAGEMENT ASSURANCES

SUMMARY OF FINANCIAL STATEMENT AUDIT

Audit Opinion—Unqualified

Restatement—No

Material Weaknesses—No

SUMMARY OF MANAGEMENT ASSURANCES

Effectiveness of Internal Control over Financial Reporting (FMFIA § 2)

Statement of Assurance—Unqualified

Material Weaknesses—No

Effectiveness of Internal Control over Operations (FMFIA § 2)

Statement of Assurance—Unqualified

Material Weaknesses—No

Conformance with Financial Management System Requirements (FMFIA § 4)

Statement of Assurance—Systems Conform to Financial Management System Requirements

Nonconformance—No

Compliance with Federal Financial Management Improvement Act (FFMIA)

	Agency	Auditor
Overall Substantial Compliance	Yes	Yes
1. Systems Requirements	Yes	Yes
2. Accounting Standards	Yes	Yes
3. United States Standard General Ledger at Transaction Level	Yes	Yes

IMPROPER PAYMENTS
INFORMATION ACT
REPORTING DETAILS

IMPROPER PAYMENTS INFORMATION ACT REPORTING DETAILS

To comply with the *Improper Payments Information Act of 2002* (IPIA) and the *Improper Payments Elimination and Recovery Act of 2010* (IPERA), the NRC incorporated improper payments testing into the 2011 A-123 Appendix A assessment.

The NRC performed a risk assessment to determine which programs would be included in the improper payments testing. According to OMB guidance, agencies were not required to review intra-governmental transactions or payments to employees. Therefore, commercial payments and grants payments remained as potential areas to test. As of March 31, 2011, total commercial payments were $113,982,097 and total grants payments were $6,932,818. OMB guidance states if gross annual improper payments exceed 2.5 percent of program outlays and $10 million of all programs or activities payments made during the fiscal year, the program or activity is susceptible to significant improper payments. Based on a risk-based analysis of susceptibility of payment streams to improper payments, management determined that the scope of the assessment would be limited to commercial payments. The scope of the assessment was further refined through the identification of 16 potential error conditions that would cause a payment to be "improper." These error conditions were grouped into three categories: payment amount, payment eligibility, and payment delivery. Test procedures were developed for each error condition.

The NRC selected a sample based on a population of the commercial payments as of May 31, 2011, that was reconciled to the general ledger. A statistician extracted a statistically valid sample of 265 commercial payments totaling $45.4 million that were divided into eight strata. This sample of 265 payments covered 3.4 percent of commercial payments and 32.7 percent of the total dollar value of commercial payments.

The results of testing identified four instances in which discounts offered by the contractor were not taken, resulting in improper payments of $3,200. Extrapolating the errors to the population resulted in $26,810 in improper payments and an improper payment rate of 0.02 percent for commercial payments in FY 2011. This rate falls well below OMB's significant improper payment threshold of 2.5 percent of program outlays and $10 million of all program or activity payments made during the fiscal year; therefore, no corrective action plans were required. As considered necessary, the NRC may review the processes in place for taking discounts.

IMPROPER PAYMENT REDUCTION OUTLOOK

Program	PY Outlays	PY IP %	PY IP $	CY Outlays	CY IP %	CY IP $	CY+1 Est. Outlays	CY+1 IP %	CY+1 IP $
Commercial Payments	NA	NA	NA	$138.6 million*	0.02%	$26,810	**	**	**

Percent and dollar value are based on a statistical sample of outlays as of May 31, 2011.

** The NRC's improper payments fall below the OMB threshold for corrective action plans; therefore, there is no need to make projections.*

RECOVERY
AUDITING
REPORTING

RECOVERY AUDITING REPORTING

The *Improper Payments Elimination and Recovery Act of 2010* (IPERA) and OMB M-11-16, Issuance of Revised Parts I and II to Appendix C of OMB Circular A-123, dated April 14, 2011, which implements the requirements of IPERA Section 2(h), require agencies to conduct payment recapture audits for each program and activity that expends $1 million or more annually if conducting such audits would be cost effective. The NRC has developed and implemented a plan to address the agency's responsibilities relative to IPERA in accordance with OMB guidance.

During a meeting with OMB that was held on July 19, 2011, the NRC discussed its plan for the NRC's recapture audit efforts, which were to be determined based on the results of the IPERA assessment. In accordance with OMB guidance, the NRC incorporated the required IPERA assessment activities into its 2011 A-123 Appendix A assessment.

In accordance with OMB guidance, the agency performed a risk-based analysis of programs and determined commercial payments to nongovernmental vendors as susceptible to improper payments, per OMB definitions. A statistical sample was selected for testing from the population of commercial payments from the beginning of FY 2011 through May 31, 2011. Based on a population of 7,734 invoices totaling $138.6 million, the

NRC selected a statistically valid sample of 265 invoices totaling $45.4 million and performed testing procedures against NRC error conditions. OMB approved this approach during the aforementioned meeting. The testing identified four instances in which discounts were not taken, resulting in improper payments of $3,200. Extrapolating the errors to the population resulted in an estimated $26,810 in improper payments and an improper payment rate of 0.02 percent for commercial payments in FY 2011.

Based on the amount of improper payments discovered, $3,200 ($26,810 extrapolated), and the approximate contractor costs of $137,205 for the IPERA testing, the NRC has determined that recovery audits are not cost effective. The NRC will devise an internal monitoring process for improper payments.

ACRONYMS AND ABBREVIATIONS

Acronym	
10 CFR	Title 10 of the Code of Federal Regulations
ABWR	Advanced Boiling Water Reactor
ADAMS	Agencywide Documents Access and Management System
ADM	Office of Administration
AGA	Association of Government Accountants
CFO	Chief Financial Officer
CFR	Code of Federal Regulations
CFS	core financial system
CoC	Certificate of Compliance
COL	Combined Operating License
CRGR	Committee to Review Generic Requirements
CSRS	Civil Service Retirement System
CUI	controlled unclassified information
DC	design certification
DHS	Department of Homeland Security
DOE	U.S. Department of Energy
DOL	U.S. Department of Labor
ECIC	Executive Committee on Internal Control
EPA	Energy Policy Act of 2005
EPR	Evolutionary Power Reactor
ESBWR	Economic Simplified Boiling-Water Reactor
FAIMIS	Financial Accounting and Integrated Management System
FBI	Federal Bureau of Investigation
FCFOP	Fuel Cycle Facility Oversight Program
FDCC	Federal Desktop Core Configuration
FECA	Federal Employees Compensation Act of 1993

Acronym	
FEMA	Federal Emergency Management Agency
FERS	Federal Employees Retirement System
FFMIA	Federal Financial Management Improvement Act of 1996
FEVS	Federal Employee Viewpoint Survey
FICA	Federal Insurance Contributions Act of 1935
FISMA	Federal Information Security Management Act of 2002
FMFIA	Federal Managers' Financial Integrity Act of 1982
FOIA	Freedom of Information Act of 1966
FR	Federal Register
FY	fiscal year
GAAP	generally accepted accounting principles
GALL	Generic Aging Lessons Learned
GPRA	Government Performance and Results Act of 1993
GSA	General Services Administration
HTGR	high-temperature gas-cooled reactor
IA	Interagency Agreements
IAEA	International Atomic Energy Agency
IAEC	Israel: Atomic Energy Commission
IG	Inspector General
IMPEP	Integrated Materials Performance Evaluation Program
Integrity Act	Federal Managers' Financial Integrity Act of 1982
IPCE	Integrated Pilot Comprehensive Exercise
IPs	inspection procedures
iPWR	Integral pressurized-water reactor

Acronym	
IRRS	Integrated Regulatory Review Service
ISA	Integrated safety analysis
ISG	interim staff guidance
ISFSI	independent spent fuel storage installation
IT	information technology
ITAAC	inspections, tests, analyses, and acceptance criteria
KINS	Korea Institute of Nuclear Safety
LWR	light water reactor
MC&A	material control and accounting
MD	management directive
MDEP	Multinational Design Evaluation Program
NEA	Nuclear Energy Agency
NEI	Nuclear Energy Institute
NGNP	Next Generation Nuclear Plant
NRC	U.S. Nuclear Regulatory Commission
NSTS	National Source Tracking System
NTEU	National Treasury Employees Union
NUREG	Nuclear Regulatory Commission document identifier
NWF	Nuclear Waste Fund
NWPA	Nuclear Waste Policy Act of 1982, as amended
OBRA-90	The Omnibus Budget Reconciliation Act of 1990
OIG	Office of the Inspector General
OMB	Office of Management and Budget
PAR	Performance and Accountability Report
POA&M	plan of action and milestones
PPS	Partnership for Public Service

Acronym	
PRA	probabilistic risk assessment
REM	Roentgen Equivalent Man
ROP	Reactor Oversight Process
RPS	reactor program system
SAPHIRE	Systems Analysis Program for Hands-On Integrated Reliability Evaluations
SCP	Secure Content Automation Protocol
SECY	Office of the Secretary of the Commission
SFFAS	Statement of Federal Financial Accounting Standards
SMR	small modular reactor
SNM	special nuclear material
SOARCA	State-of-the-Art Reactor Consequence Analyses
SRM	staff requirements memorandum
SRP	Standard Review Plan
TVA	Tennessee Valley Authority
UO$_2$	Uranium Dioxide
USAID	U.S. Agency for International Development
WFC	White Flint Complex

BIBLIOGRAPHIC DATA SHEET

NRC FORM 335 (9-2004) NRCMD 3.7 **BIBLIOGRAPHIC DATA SHEET** *(See instructions on the reverse)*	1. REPORT NUMBER (Assigned by NRC, Add Vol., Supp., Rev., and Addendum Numbers, if any.) NUREG-1542, Vol. 17

2. TITLE AND SUBTITLE U.S. Nuclear Regulatory Commission Fiscal Year 2011 Performance and Accountability Report	3. DATE REPORT PUBLISHED	
	MONTH November	YEAR 2011
	4. FIN OR GRANT NUMBER N/A	

5. AUTHOR(S) David Holley, et. al	6. TYPE OF REPORT Annual
	7. PERIOD COVERED Fiscal Year 2011

8. PERFORMING ORGANIZATION - NAME AND ADDRESS *(If NRC, provide Division, Office or Region, U. S. Nuclear Regulatory Commission, and mailing address; if contractor, provide name and mailing address)*

Division of Planning and Budget
Office of the Chief Financial Officer
U.S. Nuclear Regulatory Commission
Washington, DC 20555-0001

9. SPONSORING ORGANIZATION - NAME AND ADDRESS *(If NRC, type "Same as above", if contractor, provide NRC Division, Office or Region, U.S. Nuclear Regulatory Commission, and mailing address)*

Same as above

10. SUPPLEMENTARY NOTES

11. ABSTRACT *(200 words or less)*

The Fiscal Year 2011 Performance and Accountability Report (PAR) presents the agency's program performance and financial management information. The PAR gives the President, Congress, and the American public the opportunity to assess the agency's performance in achieving its mission and the stewardship of its resources.

12. KEY WORDS/DESCRIPTORS *(List words or phrases that will assist researchers in locating the report)* Performance and Accountability Report (PAR) Fiscal Year (FY) 2011	13. AVAILABILITY STATEMENT Unlimited
	14. SECURITY CLASSIFICATION *(This Page)* Unclassified *(This Report)* Unclassified
	15. NUMBER OF PAGES
	16. PRICE

NRC FORM 335 (9-2004)   PRINTED ON RECYCLED PAPER

AVAILABILITY OF REFERENCE MATERIALS IN NRC PUBLICATIONS

NRC REFERENCE MATERIAL

As of November 1999, you may electronically access NUREG-series publications and other NRC records at the NRC Library at http://www.nrc.gov/reading-rm.html.

Publicly released records include, to name a few, NUREG-series publications; *Federal Register* notices; applicant, licensee, and vendor documents and correspondence; NRC correspondence and internal memoranda; bulletins and information notices; inspection and investigative reports; licensee event reports; and Commission papers and their attachments.

NRC publications in the NUREG-series, NRC regulations, and Title 10, "Energy," of the *Code of Federal Regulations* may also be purchased from one of these two sources.

1. The Superintendent of Documents
   U.S. Government Printing Office
   Mail Stop SSOP
   Washington, DC 20402-0001
   Internet: bookstore.gpo.gov
   Telephone: 202-512-1800
   Fax: 202-512-2250

2. The National Technical Information Service
   Springfield, VA 22161-0002
   www.ntis.gov
   1-800-553-6847 or, locally, 703-605-6000

A single copy of each NRC draft report for comment is available free, to the extent of supply, upon written request as follows:

Address: Office of Administration,
Printing and Mail Services Branch
U.S. Nuclear Regulatory Commission
Washington, DC 20555-0001
E-mail: DISTRIBUTION@nrc.gov
Facsimile: 301-415-2289

Some publications in the NUREG-series that are posted on the NRC's Website http://www.nrc.gov/reading-rm/doc-collections/nuregs are updated periodically and may differ from the last printed version. Although references to material found on a Website bear the date the material was accessed, the material available on the date cited may subsequently be removed from the site.

NON-NRC REFERENCE MATERIAL

Documents available from public and special technical libraries include all open literature items, such as books, journal articles, and transactions, *Federal Register* notices, Federal and State legislation, and congressional reports. Such documents as theses, dissertations, foreign reports and translations, and non-NRC conference proceedings may be purchased from their sponsoring organization.

Copies of industry codes and standards used in a substantive manner in the NRC regulatory process are maintained at-

The NRC Technical Library
Two White Flint North
11545 Rockville Pike
Rockville, MD 20852-2738

These standards are available in the library for reference use by the public. Codes and standards are usually copyrighted and may be purchased from the originating organization or, if they are American National Standards, from-

American National Standards Institute
11 West 42nd Street
New York, NY 10036-8002
www.ansi.org
212-642-4900